ANTARCTICA

the blue continent

David McGonigal   Dr Lynn Woodworth

FRANCES LINCOLN

Antarctica: The Blue Continent
Frances Lincoln Ltd
4 Torriano Mews
Torriano Avenue
London NW5 2RZ
www.franceslincoln.com

First published in 2002 by
The Five Mile Press
22 Summit Road, Noble Park, VIC 3174, Australia
Tel: 61 3 9790 5000 Fax: 61 3 9790 6677
publishing@fivemile.com.au

Produced by Global Book Publishing Pty Ltd
1/181 High Street, Willoughby, NSW 2068 Australia
Tel: 61 2 9967 3100 Fax: 61 2 9967 5891

First Frances Lincoln edition 2004

ISBN 0 7112 2476 5

Printed in Hong Kong Sing Cheong Printing Co. Ltd.
Colour separation by Pica Digital Pte Ltd, Singapore

9 8 7 6 5 4 3 2 1

Captions for images in the preliminary pages
Page 1: Russian research vessel, *Academik Ioffe*
Pages 2–3: Adélie penguins
Pages 4–5: Mount Paget, South Georgia
Pages 6–7: Admiralty Ranges, Cape Adare, Victoria
Land
Page 10: Emperor penguin and chick

Details of cover images and part openers appear on
page 224.
Photographers and contributors are acknowledged
on page 224.

Publisher	Gordon Cheers
Associate publisher	Margaret Olds
Art director	Stan Lamond
Managing editor	Marie-Louise Taylor
Editors	Tracy Tucker
	Maggie Aldhamland
	Derek Barton
	Janet Healey
	Judith Simpson
Book and cover design	Stan Lamond
Cartographer	John Frith
Additional cartography	Robert Taylor
Map contour artwork	Oliver Rennert
Managing map editor	Valerie Marlborough
Map editors	Vanessa Finney
	Janet Parker
	Judith Simpson
	Jan Watson
Cartographic consultants	Vivienne Allan
	Henk Brolsma
Wildlife distribution maps	Lynn Woodworth
Principal photographer	David McGonigal
Photo library	Alan Edwards
Photo and illustration research	Gordon Cheers
	Joanna Collard
	Heather McNamara
	Tracy Tucker
Illustrations	Glen Vause
Diagrams	Mike Gorman
Index	Loretta Barnard
Typesetting	Dee Rogers
International rights	Sarah Carlon
Production	Bernard Roberts

Contents

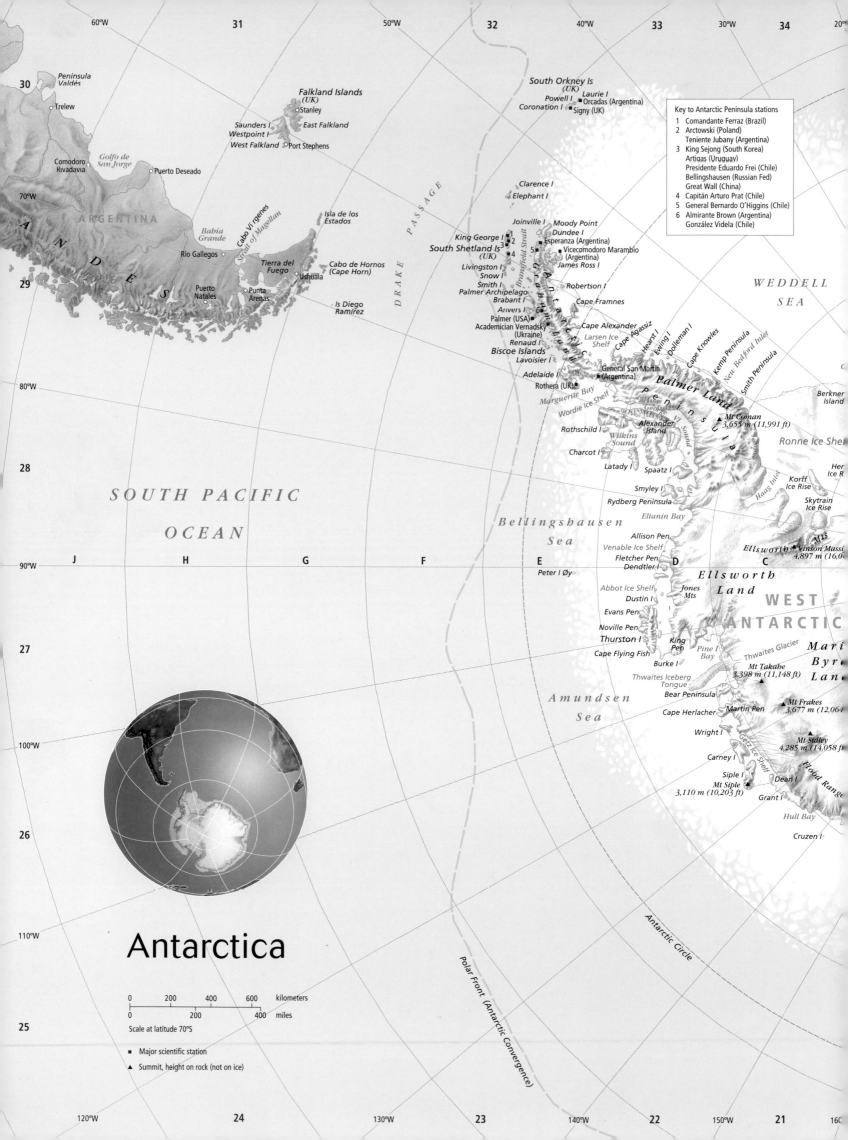

Antarctica

Scale at latitude 70°S

■ Major scientific station

▲ Summit, height on rock (not on ice)

0   200   400   600   kilometers
0   200   400   miles

Key to Antarctic Peninsula stations
1 Comandante Ferraz (Brazil)
2 Arctowski (Poland)
   Teniente Jubany (Argentina)
3 King Sejong (South Korea)
   Artigas (Uruguay)
   Presidente Eduardo Frei (Chile)
   Bellingshausen (Russian Fed)
   Great Wall (China)
4 Capitán Arturo Prat (Chile)
5 General Bernardo O'Higgins (Chile)
6 Almirante Brown (Argentina)
   González Videla (Chile)

Place names

Peninsula Valdés
Trelew
Comodoro Rivadavia
Puerto Deseado
Golfo de San Jorge
ARGENTINA
ANDES
Bahía Grande
Río Gallegos
Cabo Vírgenes
Strait of Magellan
Tierra del Fuego
Punta Arenas
Puerto Natales
Ushuaia
Cabo de Hornos (Cape Horn)
Is Diego Ramírez
Isla de los Estados

Falkland Islands (UK)
Stanley
Saunders I
Westpoint I
West Falkland
East Falkland
Port Stephens

DRAKE PASSAGE

South Orkney Is (UK)
Powell I
Coronation I
Laurie I
Orcadas (Argentina)
Signy (UK)

Clarence I
Elephant I
Joinville I
Moody Point
King George I
South Shetland Is (UK)
Dundee I
Esperanza (Argentina)
Vicecomodoro Marambio (Argentina)
James Ross I
Livingston I
Snow I
Smith I
Palmer Archipelago
Brabant I
Anvers I
Palmer (USA)
Academician Vernadsky (Ukraine)
Renaud I
Biscoe Islands
Lavoisier I
Adelaide I
Rothera (UK)
Marguerite Bay
Bransfield Strait
Graham Land
Robertson I
Cape Framnes
Larsen Ice Shelf
Cape Alexander
Cape Agassiz
Hearst I
Ewing I
Dolleman I
Cape Knowles
Kemp Peninsula
New Bedford Inlet
Smith Peninsula
General San Martin (Argentina)
Palmer Land
Mt Coman 3,655 m (11,991 ft)

WEDDELL SEA

Berkner Island
Ronne Ice Shelf
Her R

SOUTH PACIFIC OCEAN

Wordie Ice Shelf
George VI Sound
Rothschild I
Alexander Island
Wilkins Sound
Charcot I
Latady I
Spaatz I
Smyley I
Rydberg Peninsula
Eltanin Bay
Bellingshausen Sea
Haag Nunataks
Korff Ice Rise
Skytrain Ice Rise
Ellsworth
Vinson Massif 4,897 m (16,0
Mts

Allison Pen
Venable Ice Shelf
Fletcher Pen
Dendtler I
Peter I Øy
Abbot Ice Shelf
Dustin I
Evans Pen
Noville Pen
Thurston I
Cape Flying Fish
Burke I
Jones Mts
King Pen
Pine I Bay
Thwaites Glacier
Ellsworth Land
WEST ANTARCTIC
Mari Byr Lan
Mt Takahe 3,398 m (11,148 ft)

Amundsen Sea
Thwaites Iceberg Tongue
Bear Peninsula
Cape Herlacher
Martin Pen
Mt Frakes 3,677 m (12,064
Wright I
Mt Sidley 4,285 m (14,058 ft
Carney I
Getz Ice Shelf
Siple I
Mt Siple 3,110 m (10,203 ft)
Dean I
Grant I
Flood Range
Hull Bay
Cruzen I

Polar Front (Antarctic Convergence)
Antarctic Circle

Introduction

"Glittering white, shining blue, raven black, in the light of the sun the land looks like a fairy tale. Pinnacle after pinnacle, peak after peak—crevassed, wild as any land on our globe, it lies, unseen and untrodden." So wrote Roald Amundsen after discovering Antarctica's Queen Maud Range in 1911. Every visitor to Antarctica falls under the spell of the continent at the bottom of the world—a region of endless ice, strange and endearing creatures, and no indigenous inhabitants. Antarctica makes visitors feel like privileged strangers, as they would if they landed on the moon.

For every visitor to Antarctica, there are a thousand enthusiasts at home who are in thrall to the magic of this extreme environment. This has been true since the continent was first seen less than 200 years ago. This book preserves that sense of awe, but also offers a distinctly contemporary view of Antarctica. Collectively, those who have contributed have spent several lifetimes "on the ice"—yet all maintain that their passion for the southern continent remains undimmed by any uncomfortable or dangerous experiences in cold and challenging conditions. Indeed, some of the text was written on icebreakers carving a passage to the shores of the Antarctic Continent, and in tents overlooking frolicking penguins.

PART I, THE ANTARCTIC ENVIRONMENT, describes how the laws of nature and physics gave rise to these remote regions of ice and snow, and have populated them with flora and fauna perfectly adapted to survive these most hostile of conditions. It also explains how Antarctica separated from the other parts of Gondwana, and reveals how vital to the world's weather is the physical isolation imposed by the Southern Ocean.

PART II, ANTARCTIC REGIONS, gazettes the most important places in Antarctica and its surrounding oceans, from towering peaks to scientific bases, from historic sites to crowded penguin colonies and odiferous seal wallows. Many of the most memorable sites are on the Sub-Antarctic Islands and these too are detailed.

PART III, ANTARCTIC WILDLIFE, includes up-to-the-minute information about penguins, albatrosses, whales, and seals; indeed, about all the extraordinary creatures that live and thrive in Antarctica. From the simplest mosses and lichens to inquisitive penguins and giant whales, Antarctic wildlife is universally endearing in its stubborn hardiness. Antarctica is a continent dedicated to scientific study that transcends national boundaries, and recent research has added greatly to the international fund of knowledge about the unique wildlife that lives at the bottom of the world.

PART IV, ANTARCTIC EXPLORATION, relates the many chapters of a story that defies superlatives. It tells of the almost unbelievable feats of the brave men who ventured into the frozen seas at the dangerous end of the earth, first in frail wooden sailing ships, then in flimsy aircraft. The dramatic story of Antarctic exploration reflects the development of technology, although in the savage ice many innovations proved sadly inadequate.

Antarctica then, now, and in the future

The Heroic Age of Antarctic exploration coincided with the development of photography. As the true nature of Antarctica has been discovered by indefatigable human efforts, it has been invariably captured on film. Within these pages, polar landscapes and wildlife come to life through photographs spanning more than a century. Many images reproduced here were commissioned specifically for this book, and each shows the unique nature of southern latitudes that few have seen—or will ever see.

A printed page cannot capture the experience of an Antarctic blizzard's icy blast or the sensory overload generated by the cacophony of thousands of penguins. However, understanding the nature of the Antarctic illuminates a part of the world that, within living memory, consisted of little more than a blank on world maps. As the icy ramparts of Terra [once] Incognita fall to the assault of satellite transmitters, icebreakers and reliable air transport, the world is discovering regions even more marvelous than the conjectures of ancient imaginations. Within these pages we hope to contribute to an understanding of Antarctica, to inspire those who are considering venturing south and, for those fortunate enough to have visited Antarctica, to provide familiar glimpses of the world's most memorable destination.

David McGonigal   Dr Lynn Woodworth

Part I

THE ANTARCTIC ENVIRONMENT

In polar regions, the rules by which people normally live break down. These are areas of everlasting ice, days lasting six months, large populations of a few wildlife species, erratic compass readings, and the meeting of all longitudes. They are places of dazzling beauty: ice reflects rainbow shards of light, snow petrels are black beaks and eyes against the snow, and the night sky is a screen for the scintillating aurora. Here, the sciences are not dull texts but guidelines for human survival. Frozen in time and ice, Antarctica holds secrets of the earth's past—and clues to its future.

About the polar regions

From space, the earth is a blue ball capped with white at the poles, but few pause to wonder why the poles are cold enough to maintain permanent ice fields while the tropics are always hot. The energy generated by the sun is capable of heating the whole planet to 14°C (57°F), and the sun is 150 million kilometers (93 million miles) away—so why should it heat one part of the earth more than another?

The main reason is that the earth is spherical, with the equator pointing directly at the sun but the poles tilting away so that sunlight strikes the surface at an oblique angle. At 30° north or south of the equator (roughly the latitude of Florida or Sydney) this cuts the amount of sunlight received to about 86 percent of that falling on the equator. At 60° (about the latitude of Oslo or the South Sandwich Islands) the intensity of sunlight is reduced to 50 percent, and by 80° (about the latitude of the northern coast of Greenland or the edge of the Ross Ice Shelf, in the Antarctic) it has fallen to 17.4 percent. From the poles the sun is lower in the sky and its rays less warming. If the earth were not tilted, the poles themselves would receive no sunlight at all, and the sun would travel around the horizon, never rising above it.

Two other factors contribute to depriving the poles of the sun's warmth. First, the greater angle means that the sun's rays must penetrate more of the earth's atmosphere before reaching the surface. More importantly, of the solar radiation that does reach the earth's surface, the ice and snow that cover the poles reflect back at least 85 percent of it into the atmosphere.

Each 24 hours the earth rotates through 360°, so that alternate heating and cooling moderate the impact of the energy emanating from the sun; if it were not so, the side facing the sun would be too hot to sustain life, and the area in shadow too cold. The poles form the axis of this rotation, so they move very little, whereas points on the equator move at 1,670 kilometers per hour (1,037 mph) to complete a rotation in 24 hours.

The earth also travels around the sun. If the poles were perpendicular to the sun's rays, the whole earth would have 12 hours of daylight and 12 hours of darkness every day; there would be no seasons—and no easily definable year. The patterns of night and day and the seasons come about because the earth's axis of rotation inclines somewhat from the perpendicular, so that for half the year the North Pole leans toward the sun and experiences summer, while for the other half it is the South Pole that is inclined toward the sun. At the two solstices—June and December—the earth is at the points in its orbit where the North Pole or the South Pole, respectively, are most inclined toward the sun. At these times one pole is bathed in 24-hour daylight and gets more sunlight than anywhere else on earth.

Previous pages

◄ **MOUNT EREBUS**

Its summit hidden by cloud, Mount Erebus is the youngest of several volcanoes that form Ross Island, and the southern-most active volcano. Castle Rock, the snow-free knob (left) is Hut Point Peninsula's tallest point. In the foreground and middle distance, pressure ridges of jumbled sea ice push against the shoreline.

◄ **MAGNIFICENT SUNRISE**

Its image distorted by the laws of atmospheric physics, the sun rises over the Ross Ice Shelf in early summer. South of the Antarctic Circle there are some winter days when the sun never rises above the horizon and an equal number of summer days when the sun never sets below it.

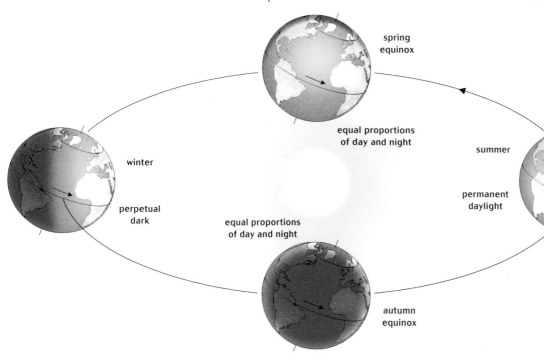

spring
equinox

equal proportions
of day and night

winter

perpetual
dark

summer

permanent
daylight

equal proportions
of day and night

autumn
equinox

The polar seasons

The planet's seasons are caused by the tilt of the earth. The closer a location is to either of the two poles, the more extreme the variation between summer and winter. As the illustration shows, during winter in the Antarctic the South Pole is tilted away from the sun, and at midwinter it lies in perpetual darkness. At midsummer in the southern hemisphere, the South Pole is tilted more toward the sun, and thus is bathed in perpetual daylight. The North Pole, being tilted in the opposite direction, experiences the opposite season, so in summertime in Antarctica it is winter at the North Pole.

➤ **RESTING ON ICE**
Midnight—but the sun is still high in the summer sky above the snow-free surface of a frozen tarn on Ross Island. Ice of such Antarctic ponds and lakes is frequently beautifully patterned. At this latitude the sun never sets between late October and late February. Continuous sunlight every day tempts many Antarctic visitors to push themselves to their physical limits.

Midway between the solstices are the September and March equinoxes, when radiation from the sun falls vertically at the equator. The name suggests that everywhere on earth has equal hours of light and darkness on that day. However, sunrise and sunset are judged not from the angle of the sun but from the appearance of its first and last rays, so different locations experience equal hours of light and darkness several days on either side of the true equinox. (The time also varies, defined by established time zones, so midday may not be midway between sunrise and sunset.) The equinoxes also mark the point where the poles graduate from six months of sunshine to six months of darkness, and vice versa. The earth does not orbit the sun in a perfect circle: when the South Pole is leaning toward the sun, the earth is about three percent closer to the sun. Nevertheless, Antarctica is much colder than the Arctic, mainly because of the dominant effect of its high polar ice sheet, but also because it is a landmass that blocks the moderating influence of the ocean.

The earth's poles

In reference to the earth, a pole is defined as either end of an axis around which the planet revolves. The earth has no fewer than seven poles. Some move over time; others are fixed, although the markers left by humans move as the Antarctic ice sheet slides toward the sea and the Arctic ice shifts with ocean currents.

Geographic poles: Fixed poles at latitudes 90°N and 90°S. Halfway between them—almost 10,000 kilometers (6,220 miles) from each—lies the equator. This is despite an attempt in the late eighteenth century to define the meter as one ten-millionth of this distance.

Poles of rotation: Movable poles forming the axis of the earth's rotation. They are within about 20 meters (66 ft) of the geographic poles, around which they rotate over 435 days. The exact locations of the geographic poles are calculated by taking the average measurement of the rotational poles.

Celestial poles: Movable poles at the positions in the sky occupied by an imaginary line extended infinitely through the geographic poles. They move because the earth wobbles on its axis over a variety of cycles that last 100,000, 41,000, or 23,000 years, due to the way in which the earth orbits around the sun.

Magnetic poles: Movable poles positioned where the magnetic field is at right angles to the earth's surface. They fluctuate daily in an oval under the influence of solar winds, moving about 2,000 kilometers (1,240 miles) in the last 400 years. Since 1841, the South Magnetic Pole has moved northwest at an average of 9 kilometers (6 miles) a year; in 2001 it was at 65°S 139°E and the North Magnetic Pole was at 79°01′N 105°08′W.

Geomagnetic poles: Calculations of where the magnetic poles would be if the earth's magnetic field worked like a simple bar magnet, they are situated at 78°39′N 69°W and 78°30′S 111°E. These poles are also the outer limits of the earth's geomagnetic field and the centers of auroral activity.

Poles of inaccessibility: In the Arctic, the position equidistant from the surrounding landmasses—84°03′N 174°51′W; in Antarctica the position, on average, that is furthest from the sea—85°50′S 65°47′E.

While not technically a pole, the point of the coldest place on earth is currently defined as Russia's base Vostok, high on the ice sheet. A temperature of −89.2°C (−128.6°F) was recorded there in July 1983.

▲ **THE SOUTH POLE**
It is said that an Archbishop came to Antarctica to say Midnight Mass at the South Pole, after having first said it at McMurdo Sound. Going to Byrd Station, he crossed the date line to say Mass again— the third time within 24 hours, always at the correct time.

> THE BIRTH OF ICEBERGS

Andvord Bay on the western side of the Antarctic Peninsula, near Anvers Island, has the spectacularly rugged landscape of mountains and glaciers that is typical of the Peninsula. These alpine glaciers occur in the high mountains and flow all the way down to the coast. When the ice flow reaches the coastline, the glaciers break off in cliffs and spawn myriad icebergs.

Polar contrasts

While the two polar regions share extremes of daylight and darkness, in other respects they differ widely.

ANTARCTIC	ARCTIC
Continent surrounded by ocean.	Ocean surrounded by continents.
South Pole: 2,836 m (9,300 ft) above sea level; bedrock 30 m (100 ft) above sea level.	North Pole: 1 m (3 ft) of sea ice; bedrock 427 m (1,400 ft) below sea level.
Deep, narrow continental shelf; restricted ice-free frozen ground; no tree line; no tundra; no native population.	Shallow, extensive continental shelf; extensive frozen ground; clear tree line; well defined tundra; circumpolar native populations.
South Polar ice sheet covering 98% of land.	Limited land ice.
Icebergs from glaciers and shelf ice measured in cubic kilometers.	Icebergs from glaciers measured in cubic meters.
Sea ice mainly annual, salty, and less than 2 m (6½ ft) thick.	Sea ice mainly multi-year, low in salinity, and more than 2 m (6½ ft) thick.
Mean annual temperature at South Pole: −50°C (−58°F).	Mean annual temperature at North Pole: −18°C (0°F).
Marine mammals (whales and seals); no terrestrial mammals.	Terrestrial mammals (reindeer, wolf, musk ox, hare, lemming, fox); marine mammals (whales, seals, polar bears).
Less than 20 bird species between latitudes 70° and 80°.	More than 100 bird species between latitudes 75° and 80°.
Lichens at latitude 82°.	About 90 flowering plant species at latitude 82°.

Measuring Antarctica

The rock and permanent ice of the Antarctic landmass cover about 14 million square kilometres (5½ million sq. miles), making it the fifth largest continent, and considerably larger than Europe. If the ice melted, Antarctica would consist of the East Antarctic continent and the archipelago of West Antarctica leading northward to the Antarctic Peninsula; it would then be the smallest continent, at about half its present size. The winter sea ice roughly doubles the effective area of Antarctica; if it were permanent, Antarctica would be third in size after Asia and Africa. At its deepest point, the dome of the polar ice sheet is 4,800 meters (15,800 ft) thick, and the South Pole stands on 2.8 kilometers (1¾ miles) of ice. The average elevation of Antarctica is 2,160 meters (7,100 ft)—Asia, the next highest, is about 1,000 meters (3,300 ft).

Defining the polar regions

The Arctic and Antarctic Circles, at about 66°33′, mark the furthest latitudes from the poles where there is at least one day when the sun does not dip below the horizon at midsummer or rise above it at midwinter. Because the earth wobbles on its axis—making its tilt from the perpendicular vary—these circles slowly shift back and forth latitudinally. This occurs over one of the periods known as the Milankovitch cycles. They can last 100,000, 41,000, and 23,000 years. At present the circles are moving closer to the poles, and, being about halfway to the turning point of a 41,000-year cycle (where the angle of the earth's axis is between 22° and 24.5°) they are moving at close to the maximum rate of about 14.5 meters (47½ ft) per year. In 1996 the polar circles were at 66°33′37″; now they lie at about 66°33′39″; they will reach 68° in about 10,000 years.

It would be convenient to define the polar regions as those lying within the polar circles, but conditions at the circles differ so much that definition on this basis is meaningless. In fact each region has several different definitions, and the criteria also differ between the south

and the north polar regions. For example, the Antarctic Peninsula extends beyond the Antarctic Circle and almost all the winter sea ice develops north of the circle.

Antarctica is defined as the Antarctic Continent, and "the Antarctic" has a variety of definitions but is loosely defined as the area south of 60°S. An alternative definition of the Antarctic is that it lies south of the natural boundary formed by the Polar Front—the convergence of oceanic waters that encircles the continent roughly between latitudes 50°S and 60°S. There is also a legal definition, embodied in the Antarctic Treaty, which defines Antarctica as everywhere south of 60°S.

The Arctic, on the other hand, is not a landmass but frozen sea surrounded by land; even in winter the sea ice stops forming well north of the Arctic Circle. One defining criterion for the Arctic region is the northern limit of the tree line: in Scandinavia there are farms and towns that lie on the Arctic Circle, and trees will survive even further north. A definition that effectively covers the same area is based on temperature: the Arctic is defined as a northern region where the average temperature during the warmest month is under 10°C (50°F). DM

◄ VOLCANIC PLUG
A black plug of volcanic rock off the north coast of Siberia rises above the snow and ice in the Kara Sea in the high Arctic. The definition of the Arctic is based on quite different parameters to those used to define the Antarctic. The Arctic is, essentially, a frozen sea, while the Antarctic includes the large continental landmass of Antarctica.

The polar landscape

Antarctica, sometimes called the Crystal Desert, has two faces today: a visible one largely of ice, and a masked one of bedrock. Ice averaging 2.3 kilometers (1½ miles) thick covers over 98 percent of the continent. In other continents, "land" is those parts lying above water; however, in Antarctica, "land" is predominantly ancient bedrock clothed in water that is crystalline except rarely in midsummer. Snow equivalent to only about 2 centimeters (¾ in) of water accumulates over much of the interior. In places, major mountain ranges 3,000 or more meters high (9,850 ft), such as the Gamburtsev Subglacial Mountains of East Antarctica, lie entirely hidden. An Antarctic "island"—the bedrock of a mountain peak or hill encircled by ice—is called by the Inuit name *nunatak*.

▲ FIRST WINTER LANDING
Cape Adare, at the northern end of Victoria Land, is a complex of overlapping, small volcanic cones. Named by Captain James Ross in 1841 for his friend Viscount Adare, MP for Glamorganshire, it was also the site of Carsten Borchgrevink's landing.

The masked face of Antarctica

The geological framework of Antarctica's approximately 14,200,000 square kilometers (5,500,000 sq. miles) can be inferred from scattered nunataks and ice-bathed mountain ranges. Extensive geophysical surveys using seismic soundings, rock magnetism, radio echo-sounding, and gravity have established the features of the buried face. Comparisons with continents to which Antarctica was joined until 130 million years ago as the hub of a giant southern continent, Gondwana, also provide useful clues.

Antarctica is roughly pear-shaped: the large, bulbous East Antarctica lies mostly in eastern longitudes; the smaller West Antarctica, with the spine of the Antarctic Peninsula stretching toward South America, is in western longitudes. The Ross Sea at the southern end of the Pacific Ocean and the Weddell Sea at the south end of the Atlantic Ocean deeply indent the continent's outline. In the nineteenth and early twentieth centuries, geologists speculated that the Ross and Weddell seas might be connected beneath the ice of the interior, but surveys during the 1957–58 International Geophysical Year (IGY) clearly showed that an ice-buried ridge above sea level links the Ellsworth Mountains of West Antarctica to the Transantarctic Mountains of East Antarctica.

A mountainous continent

Antarctica contains several of earth's major mountain belts, the largest being the great belt of the Transantarctic Mountains, a chain of nunataks, mountains, and ranges 3,200 kilometers (1,990 miles) long, extending with some interruptions, from Cape Adare on southernmost Pacific shores to Coats Land on the Atlantic side. These mountains reach to within about 500 kilometers (310 miles) of the pole, and form a topographic marker between East and West Antarctica.

The Antarctic Peninsula forms a second major mountain system that stretches about 1,600 kilometers (990 miles) south from Drake Passage to disappear as a chain of nunataks beneath the ice sheets of Ellsworth Land. An early ship's captain noted that there was a great similarity between rocks in the Peninsula and those of South America's Andes, and called the belt the Antarctandes. For example, the Andes in Chile and numerous Antarctic Peninsula rocks show evidence of copper deposits. Where this belt disappears in Ellsworth Land, its structures begin bending westward and seem to extend, largely beneath ice, to reappear along the coasts of the Bellingshausen and Amundsen

◄ ROCKS OF ANCIENT LINEAGE
Nunataks of the Framnes Mountains in the vicinity of Amery Inlet occupy an area geologically a part of the Precambrian shield of East Antarctica. They were first sighted by the BANZARE (British, Australian and New Zealand Antarctic Research Expedition) party led by Douglas Mawson in 1931.

seas. In the Cretaceous Period (144 to 65 million years ago), when ice-free Antarctica was connected to Australia and dinosaurs and other reptiles roamed, these Antarctandean mountains apparently extended the Andes around the southern Pacific to the eastern rim of Australia. Much later, this rim broke off and moved eastward during the opening and widening of the Tasman Sea to form New Zealand.

The spectacular Ellsworth Mountains at the head of the Weddell Sea, where Antarctica's highest point—Vinson Massif—rises to 5,140 meters (16,864 ft), are a geological enigma. Sandstones, shale, limestone, and ancient glacial deposits of Paleozoic age reveal a geological history more like that of East than West Antarctica. These mountains in West Antarctica contain the long-extinct plant *Glossopteris,* a late Paleozoic tree and a fossil characteristic not only of East Antarctica but of all Gondwana. Geologists think that these mountains probably originally formed somewhere in the Transantarctic Mountains and later became displaced.

Geophysically, the average thickness of the earth's crust in Antarctica matches that of other continents, roughly 30–32 kilometers (19–20 miles) thick in West Antarctica and about 40 kilometers (25 miles) thick in East Antarctica. A sharp change in thickness along the Transantarctic Mountains front may indicate a deep crustal fault system. The overall crustal stability of present-day Antarctica is confirmed by the absence of any significant earthquake activity: earthquake activity is recorded only sporadically in a few volcanoes.

▲ PEAK LANDSCAPE
Spring sunshine bathes the ice-locked coastline and Transantarctic Mountains north of Terra Nova Bay on the western shoreline of the Ross Sea. Mount Melbourne, one of Antarctica's few volcanoes considered to be "active" is near the coastline (right). The Deep Freeze Range occupies the middle distance, with the Eisenhower Range beyond it.

The formation of Antarctica

Geologists use the term tectonics (as in "plate tectonics") for movements of the earth's large, rigid, crustal plates, and for their deformation by folding and breaking (faulting) of rocks under compression, or by tension at places where, moving differently, the plates touch. Lavas—melted rocks—erupt along mid-ocean ridges as the ocean floors widen and continents separate. Along collision zones, surface rocks can be carried to great depths and mountain ranges squeezed up. At great depths, the rocks change under high pressures and temperatures, and in places melt. Lavas rise along conduits to form volcanic arcs where plates collide; the Andes, for instance, resulted from the collision of Pacific and South American plates. The geological map of Antarctica reveals a long history of plate collisions. Some geologists hypothesize that the northern extremity of Victoria Land, near Cape Adare, was an island mass that collided with Antarctica by plate-tectonic movements. Because it originated away from the continent, it is termed an "exotic terrane" of rocks.

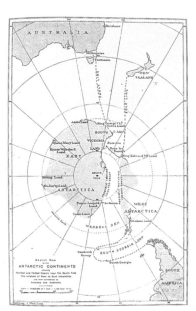

▲ A CRUCIAL QUESTION
Was Antarctica a single continent or really two, as depicted in this vintage map? That was one major question that geophysicists of the 1957–58 IGY (International Geophysical Year) wanted to answer while they were exploring the land beneath the ice by seismology.

Making mountains

Like other continents, Antarctica has a tectonic framework consisting of a long-stable core of rocks of Precambrian age (older than 543 million years) called a shield, adjoined by orogenic belts of younger rocks.

The Antarctic Peninsula formed as a volcanic arc during the Mesozoic Era, and was an active arc up to the beginning of the Cenozoic Era, about 60 million years ago. By 30 million years ago, South America stretched northward from Antarctica, became disrupted, and formed the small, new plate of the Scotia Sea.

Continents adrift

The geological history of Antarctica is largely the record of ancient plate movements. Pangaea, a supercontinent that encompassed most of earth's continental crust from more than 300 million to about 200 million years ago, rifted apart to form Laurasia, a northern landmass encompassing today's Europe, North America, Asia, and Gondwana, which later broke up into the southern continents. These ancient lands fractured along faults and split apart, perhaps many times. The Atlantic seems to have opened once, then closed, and then reopened again into today's ocean.

It is now accepted that Gondwana broke apart along a number of rifts that developed into mid-ocean ridge systems, and that new oceans carrying fragments of continents spread ever wider.

Antarctica forms one of a small number of the earth's rigid crustal plates. Today's plates formed as they broke and rifted apart along faults of some earlier plate, under crustal stresses produced by thermal plumes rising from the earth's interior. Geologists find evidence in the Ross Sea region and Marie Byrd Land for a major zone of rifting, called the West Antarctic Rift System.

East Antarctica's Precambrian shield was part of a much larger rock shield that included much of western Australia, southern Africa, Madagascar, and India. Some of the oldest rocks yet known—about 3,800 million

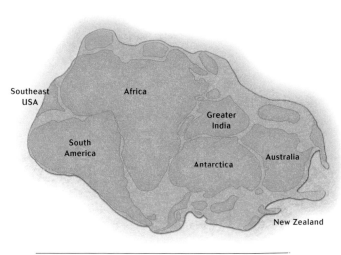

▲ GONDWANA
The southern supercontinent Gondwana, as it was about 180 million years ago. It comprised all the present-day southern continents and started to show signs of rifting at this time. A landmass consisting of Africa, South America, and India was first to separate; Australasia split off later.

ERA	PERIOD	EPOCH	YEARS AGO (millions)
Cenozoic	Quaternary	Holocene	The present
			0.01
		Pleistocene	
			1.8
	Tertiary	Pliocene	5.3
		Miocene	
			24
		Oligocene	34
		Eocene	55
		Paleocene	65
Mesozoic	Cretaceous		
			144
	Jurassic		
			206
	Triassic		
			248
Paleozoic	Permian		290
	Carboniferous		
			354
	Devonian		
			417
	Silurian		443
	Ordovician		
			490
	Cambrian		543
	Precambrian		
			3800

years old—are reported from such places. Original sedimentary and igneous rocks became deeply buried in the crust, and tectonic processes again and again transformed them by metamorphism into many types of gneiss and schist. Eventually, late in the Precambrian Era, a plate of an ancestral Pacific Ocean developed.

As it opened, this ancestral plate pushed against and collided with the adjoining continental plate of Pangaea, and was dragged beneath the continent in the process geologists call subduction. Where subducted crust was dragged sufficiently deep, parts melted under the high temperatures, forming molten silicate rock material, called magma, that was injected upward to crystallize as granite, or extruded in volcanoes. The Pacific crust collided more than once along the edge of "paleo-Antarctica," marked by today's Transantarctic Mountains. Ancient mountains formed and eroded. The processes successively added to the continent, which grew toward the Pacific Ocean.

▷ BLUE SKY, WHITE PEAKS
Towering, snowy peaks rise high above Cape Renard, which separates the Danco and Graham coasts of the west side of the Antarctic Peninsula. They were first sighted in 1898 by the men on board Adrien de Gerlache's Belgian Antarctic Expedition on the *Belgica*.

▽ A LANDSCAPE SHAPED BY ICE
The rugged western coast of the Antarctic Peninsula vividly encapsulates what a powerful force ice can be in sculpting these mountains into a landscape of valleys and hills. Valley glaciers spill over the vast cliffs into the sea and, in doing so, they spawn huge icebergs.

Fossil leaves of *Glossopteris*. Fossils of this extinct plant are found in all the fragments of the supercontinent Gondwana. These fossils were the key to proving Gondwana's former connections. When explorer Scott was found dead in his tent, *Glossopteris* fossils were found with him—having abandoned all unnecessary equipment, his group had retained the fossils, recognizing this plant's scientific significance.

Found throughout the Transantarctic Mountain range is a thick sequence of light yellow-brown sandstones called the Beacon Sandstone. Apart from being uplifted, they have barely been disturbed since their deposition over a vast period of time—from 190 to 390 million years ago.

The Beacon Sandstone

Antarctica's geological evolution followed a course that was generally similar to that of the other Gondwanan continents. Most of the record through the early Mesozoic Era, up to about 248 million years ago, is well displayed in the rock sequence of the Transantarctic Mountains, called the Beacon Sandstone by early British explorers. To the excitement of paleontologists at the British Museum, pioneer geologists of the British Scott and Shackleton expeditions in 1904–12 collected leaf fossils and coal samples of late Paleozoic age from the Beardmore Glacier at the edge of the polar plateau, and fossils of Devonian Age fish (about 380 million years old) from the quartzitic sandstones near McMurdo Sound. The coals were found to be extensive and were considered a potential mineral resource. Fossils of late Paleozoic amphibians and dinosaurs have now been found, and also of the late Paleozoic leaf, *Glossopteris*, a hallmark of Gondwana in the Permian Period. These are compelling evidence that the southern lands were once connected, and separate from the northern continents.

The Beacon Sandstone (now called the Beacon Supergroup), has since been mapped throughout the Transantarctic Mountains and at places deep within East Antarctica. It generally forms flat beds of Devonian or older to Jurassic age, overlying an eroded surface on strongly folded, trilobite-bearing Cambrian limestone and other bedrock. Such structural relations, combined with an absence of sedimentary rocks aged between Cambrian and Devonian, indicate that strong deformation and mountain building occurred after Cambrian and before Devonian time. Granites also formed during this event, which is known as the Ross Orogeny. Effects of that orogeny and of an older one of the late Precambrian Era, called the Beardmore Orogeny, are known throughout the Transantarctic Mountains. They are believed to result from collisions between an ancestral Pacific plate and the East Antarctic shield. During the late Mesozoic and early Cenozoic Eras, the continent continued to enlarge by subduction of the Pacific Ocean crust, accretion (the addition of new material), and formation of volcanic arcs along the Antarctic Peninsula and its extensions, in an event called the Andean Orogeny.

Gondwanan folding

During the earliest Mesozoic Era, rocks as young as Permian (290–248 million years) were strongly folded, affecting Antarctic regions only around the margin of the Weddell Sea, including the Antarctic Peninsula. In the middle and late Mesozoic Era, heat currents ("plumes") rising through the earth's mantle led to successive crustal rifting accompanied by volcanism on an immense scale. Gondwana was dismembered, and the new, smaller plates of today's continents formed.

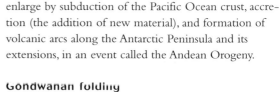

Isolation sets in

Plant and animal migration routes that had apparently freely interconnected all the southern continents were cut off at various times. Africa rifted off first; many flora and fauna species are not represented in Africa. Antarctica was becoming isolated at a time when land mammals were diversifying and flourishing elsewhere, populating all the other continents. It had long been thought that Antarctica was a migration path for marsupials moving between South America and Australia during earliest Cenozoic time, and proof of this was found in the 1982 discovery of a marsupial fossil on Seymour Island near the Antarctic Peninsula. The final phase in Antarctica's isolation was the development of a rift-fault system with Australia near the end of the Cretaceous Period.

Volcanic activity indicates the continuation of the plate-tectonic processes, as in the West Antarctic Rift System, in northern parts of the Antarctic Peninsula, and in the highly active volcanoes of the South Sandwich Islands of the Scotia volcanic arc at the eastern end of the Scotia Sea. Continuing eruptions of the volcanic caldera of Deception Island, with major activity in 1967–70 that destroyed two research stations, mark the development of a new ocean-ridge spreading system: the South Shetland Islands are moving westward from the Antarctic Peninsula and widening Bransfield Strait. Modern global positioning system (GPS) instruments are now directly measuring crustal movements and movements of continents to test past speculations by geologists and geophysicists. AF

▽ ROCK GLACIERS
A light dusting of snow highlights the rock glaciers at the head of the Beacon Valley, a Dry Valley in the Transantarctic Mountains. Composed of rock debris held together with interstitial ice, they move very slowly down the valley—at only 0.5 to 2 meters (1½ to 6½ ft) per year. The dark rocks are Jurassic basalt magma that was forced sideways, between layers of lighter sandstone, as thick horizontal sheets.

Evolution in Antarctica

▼ CLINGING TO LIFE
Plant life in Antarctica today has a tenuous existence. This whale skull is encrusted with lichen, one of the major plant forms now present in Antarctica—a far cry from the polar forests that once covered much of the continent.

Antarctic wildlife is usually perceived in terms of the animals that live in the region, either never visiting land, or using it as a temporary base for breeding. Land plants are mainly mosses and lichens, and only two flowering plants—a grass and a cushion plant that cling to the Antarctic Peninsula. However, probably as recently as 5 million years ago, woody flowering plants were present well inland, suggesting complex and diverse vegetation, and there is evidence of much more complex plant and animal communities even further back in time.

The continent now called Antarctica has occupied its present polar position for tens of millions of years, but long ago, it was nowhere near the South Pole, and even when it was, it supported life forms as complex as those of any landmass in low latitudes: for example, the *Glossopteris* flora that dominated Gondwana 248–290 million years ago. Scientists once believed that Antarctica's high latitude prevented the development of complex vegetation, because plants could not survive the extremely harsh conditions of long polar winters. However, experiments on living plants related to those that thrived in Antarctica in the remote past have proved that plants can survive such conditions, especially if temperatures are not too high. The extreme latitudes of

Antarctica are no longer seen as an impediment to the development of complex plant and animal communities.

The reduction of Antarctica's plant life to extremely sparse and simple vegetation is one of the greatest natural extinction events in the history of the earth. The cause is clear—extreme climate change. The history of Antarctic life over the past 60 million years clearly shows the impact of climate change, and the potential impact of human activity on life on earth.

Climate change

The climate of the southern hemisphere is dominated by massive oceans, their currents spanning large latitudinal belts and exerting a profound influence on climate. Water warmed by the sun in equatorial regions can transfer energy to high latitudes, but the Antarctic Circumpolar Current, which circulates vast amounts of water around Antarctica, never leaves very high latitudes, and so draws very little energy from incoming sunlight. This body of extremely cold water is the main reason for the freezing conditions that exist in Antarctica today.

When the southern continents were part of the supercontinent Gondwana, no circum-Antarctic current could form because there was no Antarctic continent; the major southern currents flowed from equatorial to polar latitudes and back again, so that the water reaching high latitudes remained relatively warm, and produced much milder conditions there; indeed, it was sometimes so warm that there was no polar ice cap. Warm seawater leads to high evaporation, and, consequently, high rainfall. In those times, land at very high latitudes was both warm and wet—perfect conditions for the development of complex forests.

The rifting of Gondwana, the separation of Australia, and then the opening of the Drake Passage between South America

and Antarctica allowed full development of the Antarctic Circumpolar Current. Circulation of this cool water forced the formation of the Antarctic ice sheet (although prior to this, higher areas may have had ice caps), which reflected much of the incoming solar radiation, making it even cooler so that the ice sheet extended even further, cooling the surrounding ocean. This feedback system is complex and often unpredictable, but the overall outcome was the modern Antarctic environment, where the ancient conditions suitable for complex forest growth are long gone.

Evidence for past Antarctic life

Fossils are usually quite common, but Antarctica is a special case, where the fossil record is particularly difficult to assemble. Today, the continent lies mostly beneath a thick ice sheet that either conceals fossils or has scoured them from the surface and deposited them in the Southern Ocean. Fortunately, the fossil records of other Gondwanan landmasses that were once connected to Antarctica—for example, South America, Australia, and New Zealand—are much more complete. They provide many ideas about Antarctic species during times when the ice cover was greatly reduced, or even absent.

When the Antarctic climate was able to support complex ecosystems, day length was a major factor affecting life. Generally, observation of similar living organisms casts light upon plant and animal communities of the past. However, all present-day high-latitude areas, with very long days in summer and continuous darkness during at least part of the winter, have extremely cold climates that prevent complex forest vegetation from developing. There are no modern forest communities growing under these light conditions, which makes reconstruction of the past more difficult.

> **SPARSE COVERAGE**
Once diverse and widespread In Antarctica, mosses now grow only sparsely on the Peninsula and along the coast of the Continent. Vigorous moss beds occasionally form, often along drainage lines, where conditions are favorable.

⌃ **FOSSIL RECORD**
Around 40 million years ago the Southern beech, *Nothofagus*, was dominant in parts of Antarctica. At least some of the species were winter deciduous and shed leaves that were trapped in underwater sediments.

◄ WINTER DECIDUOUS
Nothofagus was prominent in
Antarctica until relatively recently,
and many of the species found
in fossils shed their leaves in
winter. *Nothofagus pumilio* In
South America, shown here
during leaf turn, is a living winter
deciduous species.

▽ AUSTRALIAN ENDEMIC
Nothofagus gunnii, a Tasmanian
endemic species, is the only winter
deciduous species in Australia
today, and the only living winter
deciduous *Nothofagus* species
outside South America.

◄ ARCTIC EQUIVALENT
This fossil *Nothofagus* wood, probably five to
seven million years old from the Sirius Formation,
shows very small, asymmetrical growth rings,
suggesting horizontal growth in harsh conditions.
The growth pattern is similar to that in Dwarf
Arctic willow that survive today on the tundra.

Nothofagus

△ PERFECT FORM
A single, beautifully preserved
Nothofagus leaf from 40 mil-
lion-year-old sediments on
King George Island.

◄△ GROUND HUGGER
A single mat of fossil leaves
of *Nothofagus beardmorensis*
was discovered in Sirius Formation sediments. The plant
was probably ground hugging,
as shown in the reconstruction
on the left, with relatively
large and fragile leaves, as
seen above.

After the break up

Some prominent plants in the early Antarctic flora still exist elsewhere today, and their ecology is well understood. A good example is *Nothofagus*, the southern beech, which has been studied extensively across its modern range. In South America, New Zealand, and New Guinea, this species tends not to regenerate continuously where climatic conditions favor complex forests. In these environments, its seeds germinate and establish in fresh, cleared subsoil after catastrophic disturbance, which is common in the unstable mountain chains. This behavior is probably similar to regeneration early in the history of *Nothofagus* in Antarctica, when developing rift valleys provided unstable habitats. Until about 40 million years ago, unstable habitats at high southern latitudes provided a migration pathway for early flowering plants. They were also an opportunity for incidental populations of widespread species to become isolated, which may have been a critical factor for evolution.

The first Antarctic flowering plants—herbaceous plants or shrubby trees—did not appear until later. All these flowering plants migrated eastwards across Antarctica, from South America towards Australia and New Zealand.

In the Late Cretaceous period (99–65 million years ago), there were strong similarities between vegetation within the same latitudes of Australasia, Antarctica, and southern South America. Conifers and *Nothofagus* were the major species of the tall, open forests. There were also some Proteaceae and Myrtaceae, although both were less diverse than in Australia, where they are still prominent. By 65 million years ago, vegetation dominated by woody flowering plants was well established in Antarctica.

Major extinctions

The end of the Cretaceous period, about 65 million years ago, is noted as a time of major extinction, probably resulting from the impact of a massive extraterrestrial body. There is scant evidence of this event in Antarctica. Scientists have suggested that the collision occurred in June (mid-winter in high southern latitudes), when animals and plants, for example, winter-deciduous trees, would have been dormant. This scenario explains the paucity of evidence for the effect of the impact on Antarctic life. However, information is difficult to obtain and it is too early to conclude too much from current data.

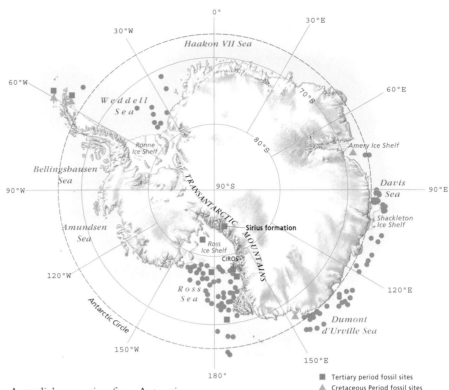

▲ **FOSSIL SITES**
Cretaceous and Cenozoic fossils are primarily located in offshore sediments. The fossils are mainly pollen and though originally laid into sediments on the Continent have been relocated (recycled) by ice scouring. For this reason, the fossil record is often difficult to interpret.

Australia's separation from Antarctica began in the Late Cretaceous period, although opportunities for plants and animals to cross water gaps may have persisted for a long time thereafter. The Drake Passage probably began to open about 23–25 million years ago, but details are lacking concerning how much dry land there was between the Antarctic Peninsula and South America before then. There are few Antarctic fossils from the past 65 million years, but some are very instructive.

Cenozoic plant macrofossils (less than 65 million years old) are known only from the Antarctic Peninsula, a drill hole in McMurdo Sound, and the Sirius Formation in the Transantarctic Mountains. Controversy surrounding these fossils centers on the accuracy of the identifications, and the age of the sediments containing them. Most are poorly preserved, which makes classifying them difficult, especially for *Nothofagus* leaves, which are relatively common.

➤ **LONG-TIME SURVIVOR**
The waratah belongs to the ancient Gondwanan plant family, Proteaceae. This family occurred in Antarctica for tens of millions of years. Today, the species occurs only in Australia.

▲ SIMILAR SITUATION
Mountain beech (*Nothofagus solandri*) occurs in New Zealand and forms part of a cool temperate ecosystem which is probably similar in structure to forests that occurred in Antarctica prior to the separation of Australia and South America. The very wet conditions are conducive to fern and moss growth on the trunks.

Pollen grains fossilize more readily than other plant parts. The pollen record, however, is often hard to interpret because much of it is not preserved where it was deposited, but lies in offshore sediments where it has been carried by ice scouring. Nevertheless, there is evidence of quite diverse and endemic species of flowering plants in the early Cenozoic era, and this vegetation persisted without much change until about 35 million years ago. Glaciers had probably reached sea level by then, so this vegetation persisted into the early phases of chilling.

The pollen record does not reveal when increasing cold and ice cover eliminated the Antarctic Cenozoic vegetation, largely because of confusion caused by the movement of the pollen once glacial erosion and sedimentation had begun. This was complicated by the discovery of probable Pliocene fossils from the Sirius Formation.

There is evidence for several glaciations on the Antarctic Peninsula between about 20 and 100 million years ago, separated by non-glacial periods when macro-fossils were deposited. During the Late Cretaceous period and the Early Cenozoic era, there were enough refugia (havens during climatic change) in Antarctica or nearby to allow woody plants to retreat to them during a glaciation, and subsequently to recolonize West Antarctica during the climatic improvements that

followed. But the formation of an Antarctic-wide ice cap about 32–33 million years ago may have brought about a major change in plant communities, and probably removed all woody plants from the Antarctic Peninsula for several million years. Antarctica's growing isolation and lengthening plant migration routes meant that many species could not return when the glaciation waned.

Thus, the Antarctic Peninsula's macrofossil record suggests a flora in decline through time. Presumably land fauna also declined, but there is no direct evidence for this. Away from the Antarctic Peninsula, a single *Nothofagus* leaf fragment, 24–29 million years old, from the CIROS-1 borehole demonstrates the presence of at least that genus at a much higher latitude, and well away from the Antarctic Peninsula. The record from the Antarctic Peninsula ends about 24 million years ago, but can be put in context with the Sirius Formation flora, which is probably Pliocene, 1.8–5.3 million years old (although some scientists believe it is much older).

Sirius Formation flora

The Sirius Formation sediments contain beautifully preserved pollen, *in situ* roots, leaves, and wood; the plants that produced these fossils must have been growing on site when the fossils were deposited. The fossil pollen is dominated by *Nothofagus*, although pollen from other flowering plants and conifers is recorded.

The fossil wood and leaves are also *Nothofagus*, and it can be assumed that they all came from the same species.

The stems are up to 1.3 centimeters (½ in) in diameter, and several are gnarled and contorted and contain branching junctions. Extremely narrow growth rings indicate extremely slow growth. In one specimen, more than 60 rings were measured along a radius of only 5 millimeters (less than ¼ in). The stems are distinctly asymmetrical, in a pattern known as "reaction wood." In flowering plants, this forms on the upper side of branches and in horizontal trunk wood. The dominance of reaction wood among Sirius samples suggests that these stems grew horizontally, like living Dwarf Arctic willows, which grow close to the ground, thus protected from freezing winds. Asymmetric growth rings are also present in prostrate forms of living *Nothofagus gunnii* from alpine Tasmania. Some fossil branches show evidence of traumatic events and scarring, and that is a common feature in living Dwarf Arctic willows.

Many hundreds of fossil leaves have been recovered, each with a very strong ribbed and creased pattern where it was folded like a fan in the bud. In living *Nothofagus* species, this indicates deciduousness and, along with the dense accumulation of leaves within a single thin layer, suggests that the fossil leaves are the result of a single, seasonal leaf fall.

Since the fossil wood and leaves from the Sirius group are found among glacial sediments, an environment similar to the present high Arctic is envisaged for the Antarctic Pliocene epoch. Pulses of glacial melt and rapid erosion, like those that occur in the Arctic spring and summer, probably damaged the tree stems and periodically retarded their growth. The vegetation probably resembled living Tasmanian alpine communities but was less diverse, with winter deciduous *Nothofagus*, other flowering plants, and a variety of conifers.

Controversy surrounding the Sirius sediments centers on interpretations of prevailing climate, particularly temperatures, based on the presence of the *Nothofagus* fossils and their nearest living relatives, and the implications for the Pliocene climate. The dwarfed growth forms of the fossil *Nothofagus* suggest that, although summer temperatures may have been around 5°C (41°F), these conditions would have lasted for only about three months during summer. For the rest of the year, temperatures would have been below freezing, and the plants would have remained dormant. The lower temperature limit would probably have been at least −15°C (5°F) to −22°C (−8°F), and possibly much colder, if snow cover had protected the dormant plants. This gives a mean annual temperature well below freezing, and perhaps in the region of −16°C (3°F).

Nothofagus-dominated vegetation must have been present in Antarctica for about 80 million years, until the time that the Sirius *Nothofagus* was deposited. This is consistent with a progressive simplification of Antarctic vegetation, possibly with some form of tundra vegetation towards the end of the process. It should be noted that the Sirius Formation fossils are well inland in Antarctica, suggesting that there may have been other, more diverse vegetation at lower sites closer to the coast.

No simple conclusions

Scarceness is the most obvious feature of the vegetation of Antarctica today. Many scientists did not believe it possible that Antarctica was once thickly vegetated. However, a combination of plate tectonics and other factors can be used to explain a very different climate at high latitudes in the past, and it is now known that diverse life was possible in Antarctica without other physical changes. It is particularly interesting that, while Antarctica is now almost barren, the rest of the southern hemisphere is covered in wildlife that owes much to this enigmatic region.

The Cenozoic decline in Antarctic plants and animals was undoubtedly climatically induced, but there is little agreement about other details. Evidence from the Antarctic Peninsula demonstrates a gradual impoverishment of the flora in response to glacial cycles, coupled with the increasing isolation of the continent and thus a strengthening of migration barriers during succeeding warmer phases. However, the probable Pliocene Sirius Formation fossils suggest that woody vegetation remained in Antarctica long after most paleoclimatologists had previously suggested that it was far too cold. The fact that one fossil find can cause so much uncertainty shows how much there is still to learn about the history of life on this mysterious continent. RH

▲ GROWTH RINGS
This flowering plant wood fossil shows well defined growth rings, indicative of a strongly seasonal climate. The cells in the wood are characteristic of flowering plants and are for water transport through the plant.

Evolutionary novelties

Evolutionary novelties are the result of major evolutionary changes in plants and animals that occur over relatively short periods of time. These changes often produce quite distinct life forms, distinguished as genera or even families of species that have many characteristics in common. High latitude areas were, and still are, an important source of evolutionary novelties. The reason for this is not well understood, although several theories have been proposed. However, these theories are difficult to test.

The Weddellian Biogeographic Province, extending from southern South America along the Antarctic Peninsula and West Antarctica to Tasmania, southeastern Australia, and New Zealand, was the center of origin and diversification for many organisms during the past 100 million years. The development of marsupials is an obvious example. Many new plant species arose and survived in this region, including important groups like *Nothofagus*, the casuarinas, and the Proteaceae. Much of Antarctica's endemic plant and animal life remained isolated because of geographical, climatic, or biological barriers. However, many plants and animals spread northward from the Weddellian Province to mid- and low-latitude regions, where their descendants still occur today, long after their place of origin has become too cold to support such life forms.

▽ Fire oak (*Casuarina cunninghamiana*).

The Polar Front

Early explorers seeking a great southern continent found that both the air and the sea became cooler as they sailed south across the Southern Ocean. The whalers and sealers who followed them in quest of quick and easy profits noticed this too. These intrepid travelers also observed that strong currents set their ships to the east where the temperature change occurred. This transition between warm subtropical waters and cold polar waters became known as the Antarctic Convergence. Today, it is called the Polar Front.

Defining the Polar Front

The Front encircles Antarctica between latitudes 40°S and 60°S. Changes in sea temperature across the Front occur in a few sharp jumps—fronts—rather than as a gradual decrease across the width of the Southern Ocean. The two most important fronts are the Subantarctic and Polar fronts. The rapid temperature changes across the fronts are linked to the strong eastward flow of the Antarctic Circumpolar Current.

These fronts act as boundaries that define zones with different temperatures, salinities, and nutrient concentrations. The different zones also tend to be populated by distinct communities of plants and animals; early oceanographers could determine which side of the Polar Front they were on by the presence or absence of particular species of krill.

Antarctica's Circumpolar Current stretches for more than 20,000 kilometers (12,400 miles) around Antarctica. The surface speed of the current is modest, but its great depth and width make it the largest of all currents in the world's oceans. It carries about 135 million cubic meters (4,800 million cubic feet) of water per second from west to east around the Antarctic Continent— equivalent to about 135 times the flow of all the world's rivers combined.

The massive flow of the Circumpolar Current is driven by some of the strongest winds on earth. The persistent westerly winds, which are punctuated by frequent violent storms, prompted sailors to dub the southern latitudes the Roaring Forties and the Furious Fifties. The strong winds acting over the circumpolar extent of the Southern Ocean also create the largest waves on the planet.

Ocean conveyor belt

The Antarctic Circumpolar Current plays a unique role in the earth's climate system. Each of the world's major ocean basins is enclosed by land except at its southern boundary, and the Circumpolar Current functions as a pipe connecting these basins, smoothing out variations in water properties between the basins, and—more importantly—permitting a truly global ocean circulation pattern. Water, made cold and saline at high latitudes, becomes heavy enough to sink into the deep ocean;

SOUTHERN OCEAN CURRENTS
The currents of the Southern Ocean circulate west to east, unimpeded by land. This is the only place on the planet where the oceans can circulate around the globe, uninterrupted by any continental landmass. These currents loosely follow a route tracing deeper waters, and allow water transfer between oceans.

Ocean depth less than 3,500 meters (11,480 ft)

Antarctic Circumpolar Current

Antarctic Intermediate Water

Antarctic Bottom Water

warm water flows into the high latitude regions to replace the sinking dense water. The exchange of warm and cold water carries heat from low latitudes to high latitudes, cooling the former and warming the latter, and so maintaining the earth's climate. The Circumpolar Current, by allowing the free exchange of water between the ocean basins, is a key link in this so-called "overturning circulation," or "ocean conveyor belt."

Polar differences

The southern polar region differs in important ways from the northern. Antarctica is a continent surrounded by the Southern Ocean, whereas the Arctic is virtually a landlocked sea; only relatively small amounts of water are exchanged through the narrow straits between the Arctic seas and the northern Pacific and Atlantic oceans.

The two polar regions also differ in the character of the sea ice. Antarctic sea ice tends to be carried away from the continent by winds and currents, whereas Arctic sea ice is trapped by surrounding landmasses and has a tendency to pile up, becoming much thicker than the sea ice of Antarctica. Freezing of the ocean surface in the Antarctic effectively doubles the size of the continent in winter, but in summer most of this ice melts, whereas the Arctic is ice-covered all year round.

As a result of these differences, there is no Arctic equivalent of the Antarctic Polar Front. In the northern hemisphere the transition between polar and subtropical waters occurs further south, outside of the Arctic, and is not circumpolar due to the presence of landmasses. SR

▲ FORCE 11 GALE
The gale force winds and huge seas battled by the early explorers are the same as those that confront visitors to the Southern Ocean today. The ships may have changed, but the Southern Ocean's power to generate the wildest conditions on earth still poses a great challenge to modern-day sailors.

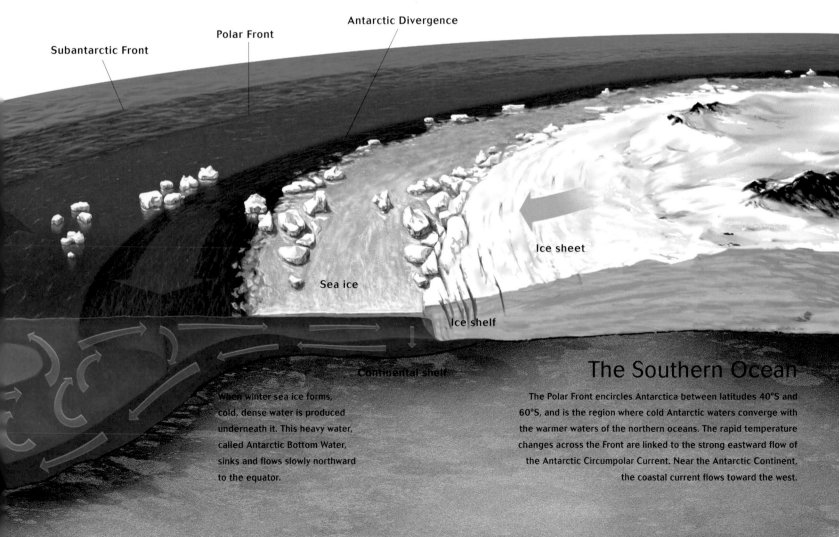

Subantarctic Front

Polar Front

Antarctic Divergence

Ice sheet

Sea ice

Ice shelf

Continental shelf

When winter sea ice forms, cold, dense water is produced underneath it. This heavy water, called Antarctic Bottom Water, sinks and flows slowly northward to the equator.

The Southern Ocean

The Polar Front encircles Antarctica between latitudes 40°S and 60°S, and is the region where cold Antarctic waters converge with the warmer waters of the northern oceans. The rapid temperature changes across the Front are linked to the strong eastward flow of the Antarctic Circumpolar Current. Near the Antarctic Continent, the coastal current flows toward the west.

Cold, dry, and windy

The north and south polar regions are "heat sinks" that influence the whole world's climate. A complex meteorological system is created by the different levels of energy received by the poles and the tropics, and is greatly affected by the spinning of the globe. Air heated at the equator rises and flows toward the poles, where it cools and sinks; this dense, cool polar air flows back to the low-pressure area created at the equator by the rising warm air.

Although the Antarctic polar ice sheet is crowned by constant high pressure, it is surrounded by a region of low pressure. Since Antarctica and South America separated, probably about 25–30 million years ago, the Southern Ocean has completely encircled Antarctica, allowing the winds to flow unimpeded. Here, eastward-bound low-pressure systems are generated in never-ending succession, circling Antarctica like a procession of spinning tops. Sailors call these latitudes the Roaring Forties, the Furious Fifties, and the Screaming Sixties.

Weather observations, first by explorers and then by scientists, have been invaluable sources of meteorological information about Antarctica. The International Geophysical Year (IGY) of 1957–58 led to the establishment of many Antarctic research bases, and a few years later the development of weather satellite technology provided a wealth of information. Today, about a hundred ground stations, some attended (mainly coastal) and some automatic (mainly across the interior), provide weather information for distribution around the world.

More than 97 percent of Antarctica is covered with snow, and this has a significant impact on climate all over the world. Solar energy is reflected from the earth in an effect known as albedo. The Antarctic surface absorbs radiation for a short midsummer period only—for the rest of the year it re-radiates more energy than it receives, the balance being restored by heat transfer from the tropics.

World's coldest

The lowest temperature recorded on earth was taken by A. Budretski on 21 July 1983 near Russia's Vostok station: it was –89.2°C (–128.6°F). Antarctica rapidly becomes colder in autumn, reaching its extreme in the lightless depths of winter. Inland, winter temperatures can range from –40°C to –70°C (–40°F to –94°F); coastal temperatures in winter range from –15°C to –35°C (5°F to –31°F).

SCULPTURES IN SNOW

Light winds deposit snow in the lee of obstacles and undulations across the polar landscape, and then blizzards erode the snow into sculptured shapes known as sastrugi. These polished forms can stand several meters high and be as hard as concrete. A vast field of sastrugi, like the one pictured here, appears as a snap-frozen, storm-tossed sea.

The Antarctic Peninsula is by far the warmest part of Antarctica; in midsummer, temperatures there can reach 15°C (59°F), while in East Antarctica they range from 0°C (32°F) on the coast to –25°C (–13°F) inland.

World's driest

The air over Antarctica is generally too cold to hold water vapor, so there is very little precipitation. Antarctica, the world's driest desert, ranks with the Sahara as the world's largest desert. But Antarctica retains what moisture it receives; about 75 percent of the world's fresh water is stored as ice, and 90 percent of that ice is in Antarctica.

World's windiest

Antarctica is the windiest place on earth. A variety of different winds blow there, from the inversion winds at the Pole to winds funneled between islands along the coast, and violent blizzards can develop with incredible speed. The wind creates its own landforms: sastrugi, irregular ridges up to 1 meter (3 ft) high, carved by blowing snow. Like sand dunes, their undulating surfaces indicate the prevailing wind direction. The strongest winds, however, are on the coast, where wind speeds of up to 300 kilometers per hour (185 mph)—twice the velocity of hurricane-force winds—have been recorded.

On 8 January 1911, Douglas Mawson's Australasian Antarctic Expedition arrived at Cape Denison. That afternoon, the wind rose as they unloaded their supplies. They waited for it to drop, but it never really did. Mawson called his account of the 1911–14 expedition *The Home of the Blizzard*—not without good reason. He had built his hut where blasts of katabatic wind flow down from the Polar Plateau. Simply put, in Antarctica, katabatic wind is cold and dense air that pours down the

COLD AND WINDY

Temperature and wind speed vary greatly over the Antarctic Continent. In general, the heart of the Continent (Dome C) is very cold, especially during the dark of winter. Antarctica's coastal regions (Cape Denison, Butler Island) are milder, warmed by the oceans, but are often the windiest areas, buffeted by freezing blasts of katabatic winds from the icy slopes.

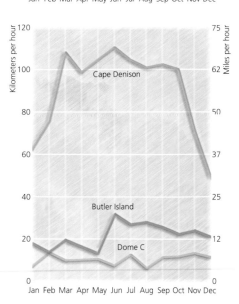

ice slope to the sea, becoming denser and picking up speed as it goes. On average, the wind speed at Cape Denison reaches hurricane force every three days. Mawson's own words sum up the human response to Antarctic weather: "… the drift is hurled, screaming through space at a hundred miles an hour, and the temperature is below zero Fahrenheit … We stumble and struggle through the Stygian gloom; the merciless blast—an incubus of vengeance—stabs, buffets and freezes, the stinging drift blinds and chokes … We had found an accursed country." DM

> READING THE PULSE

A field surveyor aligns an automatic weather station mast north, as a baseline for wind direction measurements and to maximize power output from a solar panel. Hourly meteorological readings are automatically transmitted around the world via satellite.

The Antarctic ice sheet

Antarctica's ice sheet is immense. It covers about 14 million square kilometers (5,500,000 sq. miles), and averages about 2,300 meters (7,500 ft) thick (from bedrock), with local thicknesses up to nearly 5,000 meters (16,400 ft). Antarctica's ice potentially locks up about 75 percent of the world's fresh water. Surprisingly, seismic soundings have recently confirmed the presence of many large lakes under East Antarctica, but little is known about them, especially about their age. Just above one of the largest of these lakes, Lake Vostok, the drilling of an ice core more than 2 kilometers (1¼ miles) long has been discontinued until techniques can be developed for sampling its waters without running the risk of contaminating them. It is believed that the lake's waters may contain extremely ancient organisms (perhaps now extinct elsewhere), as well as a record of climatic conditions in the world tens or hundreds of thousands of years ago.

⌃ DRILLING RIG
Law Dome is an isolated ice cap near Casey station. Due to the underlying bedrock, the ice of the dome is moving radially out from its center. Separated from the main ice sheet and circular in shape, the dome is a miniature of the Antarctic ice sheet. At the top of the dome, scientists drill ice core samples to learn about climate change, especially about the makeup of the earth's past atmosphere.

Antarctica's ice sheet

The East Antarctic Ice Sheet contains a great volume of ice, and much, though not all, of it is grounded on bedrock near or well above sea level. In contrast, much of the far smaller West Antarctic Ice Sheet lies on rock below sea level, which will be an important factor if climates warm sufficiently for the polar ice sheets to melt. Glaciologists debate whether the Antarctic ice sheets are expanding or decaying, but they generally believe that the continental ice sheets, especially that of East Antarctica, are more or less in a state of equilibrium, in contrast to some of Antarctica's ice shelves, which seem to be breaking up. The ice sheets are like settling tanks, trapping and accumulating materials from the atmos-

phere. They contain a record of ancient climates, and of global volcanic activity extending back for thousands of years, as well as trapped meteorite fragments.

Storm tracks seldom reach the interior of East Antarctica, and much of this region is true desert, where snow accumulates at very low rates, generally about 2 centimeters (¾ in) of equivalent water per year. As snow layers become buried, they transform into ice, brittle at upper levels, but becoming more plastic under pressure at deeper levels. Under the thickest areas, the ice can reach its pressure melting point and water can form, allowing basal sliding—the rapid movement of ice over bedrock under the pull of gravity. The highest part of the ice sheet, about 4,000 meters (13,000 ft) above sea level, lies not at the South Pole but in the middle of the East Antarctic Ice Sheet at about 80°S latitude and 75°E longitude, south of the Amery Ice Shelf. From near there, ice flows outward more or less radially by flow under gravity. Where it reaches the sea, much of the ice calves off as icebergs, and flows onto and adds to nearby masses of floating ice shelves that fringe the coasts. In places, ice movement is channeled into rapidly flowing ice streams. Regional ice "domes" occur at various places.

The East Antarctic Ice Sheet also flows toward and becomes largely dammed by the Transantarctic Mountains. In many places the dam is breached by outlet glaciers that squeeze through the mountains and plunge down to feed the Ross Ice Shelf, or flow into the Ross Sea, farther north. The giant, highly crevassed

A cross section of Antarctica

The shape of Antarctica, with the exception of the Transantarctic Mountains and a few locations around the coast, is lost under a blanket of ice. In some locations in East Antarctica, the ice is up to several kilometers thick.

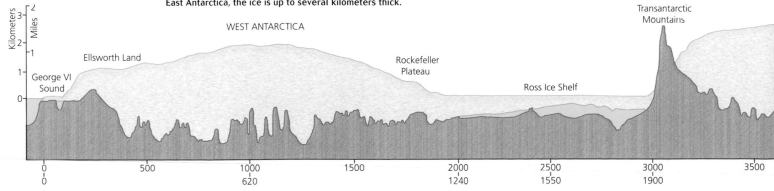

EAST ANTARCTICA

Dome C

Cape
Poinsett

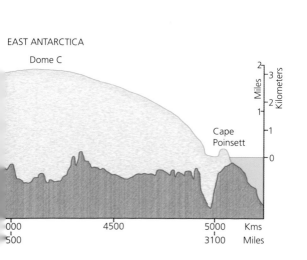

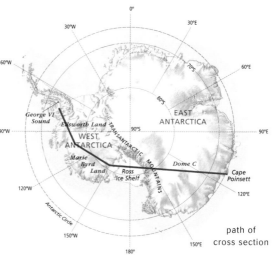

path of
cross section

▲ **WATERFALLS OF ICE**
At the head of the Taylor
Glacier in the Dry Valleys,
the ice of the Polar Plateau
pours over the hidden lip of
the Transantarctic Mountains,
(here buried under the ice).
Looking like a giant waterfall
in slow motion, the ice breaks
up into heavily crevassed
"icefalls," the tops of which
are shown here.

Beardmore Glacier was used in the early 1900s by Ernest Shackleton and Robert Scott's expeditions to the Polar Plateau on their way to the South Pole. The Axel Heiberg Glacier gave Roald Amundsen a more direct route to reach the Pole first—in November 1911.

The West Antarctic Ice Sheet forms several major ice streams that feed the head area of the great Ross Ice Shelf, and are its main source. Its center is at only half the elevation of that of the East Antarctic Ice Sheet. The accumulation rate is much higher than that of East Antarctica because many maritime storm tracks cross the area. The Antarctic Peninsula lies outside the area of the ice sheets, and it contains mostly glaciers of alpine type, as well as larger ice fields.

Theories of origin

Today we live in an ice age world (often referred to as an "interglacial" stage), with large ice sheets covering Antarctica and Greenland. Antarctica's ice sheet gives the best picture of how northern North America must have looked 20,000 years ago, at the end of the last glacial maximum, under the Laurentide Ice Sheet, which had largely disappeared by 10,000 years ago. In Antarctica, ice seems to have begun forming between 35 and 50 million years ago and the ice sheet probably remains in much the same form as it has throughout the ice age. Major glaciations seem not to be synchronous between hemispheres (though there is some debate about this), whereas the advances and retreats of smaller valley glaciers tend to be more sensitive to both global and regional climate changes.

The development of Antarctica's ice sheet seems to have commenced about 10 million years after the continent "drifted" southward into the polar region as part of Gondwana about 200 million years ago. The reason for the long delay is unknown, but oceanographers suggest that it may be due to a decline in ocean temperatures about 50 million years ago, at a time when glaciers seem to have started forming in Antarctica, probably in high country such as the Ellsworth Mountains and the Transantarctic Mountains. As climates cooled, valley glaciers advanced and coalesced into piedmont glaciers in plains at the foot of mountains, which in turn expanded, and eventually built up into ice sheets that covered the continent. The record of ice build-up is found in surrounding ocean sediments. Cores drilled from the Ross Sea in the 1970s recovered boulders and pebbles that must have been carried along by icebergs and dropped into seafloor mud about 30–40 million years ago. Such rocks, deposited far from their place of origin, are called glacial erratics. The volume of south polar ice must have fluctuated greatly since the birth of the Antarctic ice sheets, and the continent has probably been continuously glaciated since their formation, but there is some evidence of extensive deglaciation about 5 million years ago.

Evidence and opinions that try to explain the formation of glaciers and their oscillations often conflict. In 1920 the Serbian mathematician Milutin Milankovitch found evidence for cyclic variations (of 23,000, 41,000, and 100,000 years) in the earth's orbit around the sun, and in movements of the earth's axis that might explain variations in solar radiation absorbed to warm or cool the earth. Scientists of many specialties are refining the Milankovitch Cycles in ever-increasing detail in order to improve correlations with geologists' evidence from field work. Others see the sun as the prime mover of climatic change, and search for variations in solar cycles (such as the 11-year sunspot cycle), and in radiation from a "turbulent sun." These sorts of possibilities for explaining continental glaciations are not yet well understood. AF

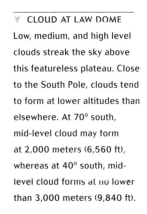

▼ CLOUD AT LAW DOME
Low, medium, and high level clouds streak the sky above this featureless plateau. Close to the South Pole, clouds tend to form at lower altitudes than elsewhere. At 70° south, mid-level cloud may form at 2,000 meters (6,560 ft), whereas at 40° south, mid-level cloud forms at no lower than 3,000 meters (9,840 ft).

➤ GLACIAL ERRATICS
Millions of years ago, these banded sandstone boulders were picked up and carried many miles in the ice before being deposited by the Vanderford Glacier on this smooth granite surface in East Antarctica.

Ice shelves and glaciers

Australian geologist Douglas Mawson may have been the first person (in 1912) to use the term "ice shelves" to describe sheets of very thick, mostly floating ice. Ice shelves make up nearly 50 percent of the Antarctic coastline. They are fed by glaciers or ice streams from Antarctica's continental ice sheets, and are additionally nourished by snow accumulating at their surface, and probably, in some cases, by seawater freezing at their base. Some areas may be aground. The Ross Ice Shelf is the world's largest ice shelf, averaging about 330 meters (1,100 ft) thick and increasing to a maximum of about 700 meters (2,300 ft) toward its southern boundary.

Ice shelves
Ice shelves generally fill embayments, and so are landlocked on three sides. They have level or gently undulating surfaces, and their seaward side spawns the tabular bergs characteristic of the Southern Ocean. Tabular bergs are usually easily distinguished from bergs formed by a calving glacier; the typical shape of a newly calved glacier berg is irregular.

Where ice shelves are joined to land, immense cracks or crevasses can develop due to ocean tides and currents. The Grand Chasm, at the head of the Filchner Ice Shelf, is such a zone of almost impenetrable crevasses. It was the first challenge for English geologist Vivian Fuchs and his companions on their way to the continent itself in the first successful continental crossing from the Weddell Sea to the Ross Sea in 1957–58. There had been two previous attempts, by German Wilhelm Filchner in 1911–12, and by Englishman Ernest Shackleton in 1914–16. Both had their ships imprisoned by the pack ice.

Most Antarctic ice shelves are small, but the three principal ones, in decreasing order of size, are the Ross,

▼ WAVES OF ICE
As the vast Ross Ice Shelf approaches immovable Hut Point Peninsula on Ross Island, it buckles, looking like huge waves about to break on land. The flat ice in the foreground is sea ice just a few meters thick that later in summer will probably break out, leaving open sea before lower autumn temperatures freeze it again. The long, straight line in the distance is a service road.

the Ronne, and the Amery. The Ross Ice Shelf lies at the head of the Ross Sea and covers an area about the size of France. Another, the Larsen Ice Shelf of the Antarctic Peninsula, has decreased greatly in recent years and seems to be breaking up. The Larsen Ice Shelf is the probable source of many tabular bergs seen by visitors on cruise ships.

The first humans to reach the Ross Sea were awed by the impressive cliff front of the Ross Ice Shelf, which to the south rises up to 50 meters (165 ft) above the sea—it seemed such an impediment to travel inland that it became known as the Ross (or Great) Barrier.

Visitors to the Ross Sea and McMurdo Station are likely to see some of the most developed pressure ridges in Antarctica. At this point, the Ross Ice Shelf, moving northward, collides with Ross Island and is thrown into spectacular giant folds.

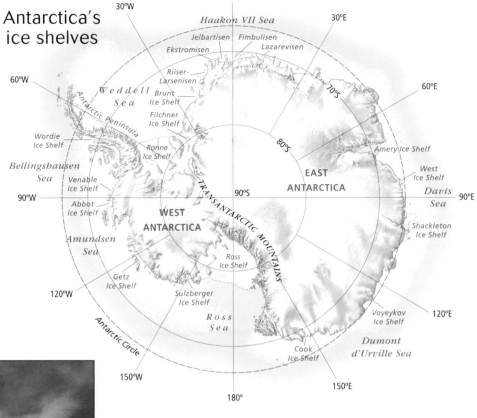

Antarctica's ice shelves

◀ **PRESSURE RIDGES**
Storms can put pack ice under immense pressure, crushing and upending floes, rafting one on top of another, and forming pressure ridges (as shown here) where the floe buckles. In some places, the resulting sea ice can be up to several meters thick.

▼ **DIRTY ICE**
Ice shelves are typically much "cleaner" than this area, called the Dirty Ice, which lies at the edge of the ice shelf at McMurdo Sound in midsummer when the surrounding sea ice has broken out. Each of the ponds in this area has a unique chemistry and is full of cyano-bacterial microbial mats.

Glaciers

In Antarctica, rivers of ice, called valley glaciers, form at the outlets of the polar ice sheets that flow down through the Transantarctic Mountains to feed the Ross Ice Shelf. The Lambert Glacier, 40 kilometers (25 miles) wide, and probably the world's largest glacier by volume, drains a major area of the East Antarctic Ice Sheet to feed the Amery Ice Shelf. The fastest-flowing Antarctic glacier known, the Shirase in Queen Maud Land, flows at a rate of 2 kilometers (1¼ miles) per year. On the Antarctic Peninsula there is no ice sheet, so glaciers are defined by the area's mountain topography. Only glaciers of alpine type are seen, including cirque glaciers that occupy high mountain amphitheaters, and valley glaciers, which flow down mountain chasms. Many of these glaciers show evidence of retreat in recent years. Glaciers can carry rocks on their surfaces and deposit them along their path, where they end up as ridges of bouldery material called moraines. Such ridges show the former extents of glaciers long after the ice has melted back.   AF

▼ SEA CAVES

When the ocean washes up against icebergs, it exploits any areas of weakness. Eventually the weakest section of ice gives way, leaving a cave-like hollow. An average berg has a life of four years. As they age, bergs split along lines of natural weakness and roll over when they become unstable.

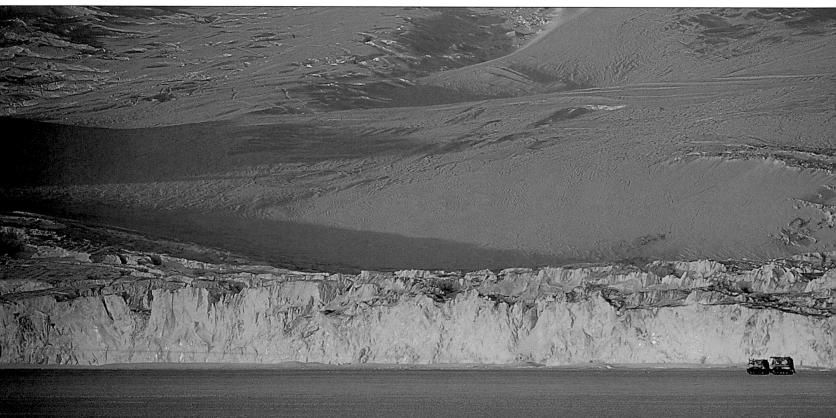

VALLEY GLACIERS

Antarctica is drained by vast systems of glaciers and ice streams. Valley glaciers, the most spectacular of them, pass through exposed mountain ranges, flowing around large, exposed mountain tops or hills called nunataks.

GLACIER TONGUE

Vast floating glacier tongues such as that of the Vanderford Glacier (seen here), can spawn large tabular icebergs not easily distinguished from true ice shelf bergs. Glacier tongues project out into the ocean, landlocked at the rear, where the feed glacier comes from; ice shelves are usually land-locked on three sides.

FROM THE MOUNTAINS TO THE SEA

The lower slopes of Ross Island are shaped by heavily crevassed glacial ice, formed by the compaction of accumulated snow. Here, at the sea edge, flat sea ice meets the glacial ice. Unbroken sea ice attached to a coastline is called fast ice, which is strong enough to support ice vehicles, and even to land large aircraft on.

Icebergs

An iceberg adrift is the archetypal image of ice in Antarctic waters. However, icebergs are not composed of frozen salt water; they are born on land and made of fresh water. They begin as snow falling on the continental ice sheet; these accumulate and are compressed over many years, as the ice sheet flows toward the sea. Upon reaching the sea, the ice spills into the water when a glacier calves, or floats on the ocean surface as an ice tongue or a massive ice shelf. The Ross Ice Shelf is just a floating ice block, abutting the fast ice along its seaward front so that it rises and falls with the tides.

Every year, thousands of icebergs break off Antarctica, scattering across the Southern Ocean, mainly below the Polar Front, where the water is cold enough to slow their melting. There are fewer icebergs in the Arctic because there is less glaciated land.

Huge tabular bergs sometimes break off ice shelves, and drift northward under the influence of winds and currents, to melt eventually in the warmer water north of the Polar Front. On rare occasions, they have been found as far north as 45°S in the South Pacific, and at 35°S in the Indian and South Atlantic oceans.

As well as the huge icebergs that regularly make the headlines, there are innumerable smaller ones. One survey carried out in 1985 found 30,000 icebergs in an area of 4,000 square kilometers (1,500 sq. miles) of ocean. Today, satellite imaging and satellite tracking gather a more accurate record.

Profiles of bergs

Great hunks of ice shelves break off into icebergs. In 2000, when the Ross Ice Shelf calved the biggest iceberg ever seen, the larger of the two segments that broke away from the shelf was about 300 kilometers long and 40 kilometers wide (185 miles by 25 miles) and had an area of more than 10,915 square kilometers (4,214 sq. miles).

Whereas ice shelves give birth to tabular icebergs, most glacial tongues (unless they are huge) create icebergs of irregular shapes that can inspire flights of fantasy. Frank Worsley, captain of Shackleton's *Endurance*, described the ice that threatened their lifeboat, *James Caird*, as they struggled to reach South Georgia: "Swans of weird shape pecked at our planks, a gondola steered by a giraffe ran foul of us, which much amused a duck sitting on a crocodile's head … "

Icebergs range from crystal clear and pure white to green, brown, and blue—even pink from algae. Most have fissures of bright electric blue and many are fringed by icicles. The shapes are unimaginable—from the hewn straight sides of a recently calved tabular iceberg to the grottoes, pinnacles, and arches of a berg that has rolled to reveal its partially melted underside.

Dangers and difficulties

The common expression—"just the tip of the iceberg"—takes on new significance in polar regions. Less than 10 percent of an iceberg is visible above the water, but the actual percentage can change depending on such factors as impurities in the ice and the salinity and temperature of the water. Although they look huge and unchanging, icebergs are inherently unstable, as they are constantly melting from both the top and the bottom.

Smaller icebergs (relative terminology if they are seen from a ship) become top-heavy as they melt underwater, and can roll quickly when their equilibrium shifts. Large bergs may seem to explode violently, as they collapse into many pieces, creating large waves and dangerous vortices.

Ice shelves spawn tabular bergs that may tip over to become capsized bergs, or they may create castellated bergs after a lot of the ice has melted from underneath. Glacial ice breaks into the ocean as smaller icebergs that decay to become "bergy bits," then "growlers," when they have melted enough so that very little is visible

▼ TABULAR BERG
Icebergs are categorized according to their shapes and sizes. This iceberg is known as a tabular berg, with its table-like top and sheer sides. Tabular bergs can be enormous, sometimes hundreds of square kilometers in surface area. What appears above the water is just a fraction of the overall size of the iceberg. About four-fifths of an average iceberg is submerged.

▶ A TYPICAL MIXTURE
A typical scene off the coast of the Antarctic: a mix of sea ice and land ice. Ice floes are sections of sea ice and can be up to 2 kilometers (1 ¼ miles) long. In between the floes are smaller tabular bergs, capsized bergs, and the beginnings of a castellated berg.

above the water. Growlers are very dangerous because they may float unnoticed by radar or helmsman until a ship collides with one. The most notable shipping disaster involving an iceberg was the *Titanic*, which sank in 1912, on its maiden voyage from Southampton to New York. The *Titanic* didn't hit a growler but came to grief on the extended underwater "foot" of an iceberg.

While most icebergs simply drift toward warmer, temperate waters, where they quickly melt, some run aground on shallow shoals and stay intact for years. In the 1970s, there was considerable interest in the idea of towing icebergs towards the desert coastlines of Australia, South America, and Africa, where their fresh water could be used to irrigate barren land. This bold plan was not completely impractical, although the cost would have been enormous, but it was abandoned after consideration of all the details. One issue was, of course, safety. The main problem, however, was the underwater bulk of the iceberg; it would almost certainly have become grounded on the continental shelf at a distance too far to pipe the water ashore.   DM

▲ BLUE ICE

As snow falls, its weight compresses the snow beneath, and it turns to ice. At first, the ice is air-filled and granular, but by the time it sinks to 50 meters (165 ft), it has become solid ice clouded with bubbles of air. As it descends deeper, the bubbles are compressed out, and the ice becomes clear blue. When a glacier containing this blue ice breaks off and floats away, it becomes a "blue berg"—one of Antarctica's most attractive features.

The frozen seas

As winter approaches, the shores of Antarctica go through a transformation. By March, after weeks of 24-hour sunlight, most of the sea ice that normally surrounds the continent has melted. All that is left is a rim of ice in some places, most notably the Weddell Sea and the Bellingshausen Sea.

The formation of sea ice

Each evening as winter draws closer, the sun dips further below the horizon and the air temperature falls enough to create a thin sheet of ice on the water's surface, seen first as ice needles and tiny ice plates known as frazil. It develops into a thin sludge that sailors call grease ice. At first the ice crystals are about 2.5 centimeters (1 in) wide and about 0.15 centimeters (less than ⅛ in) thick, but when the sea is calm, they can grow quickly to form sheets, especially when snow falls on top of this skin of ice. The plastic ice, up to 10 centimeters (4 in) thick, which bends readily with the waves, is called nilas. Wind and wave motion, however, may break the ice into plates, and then bang them together to create pancake ice—a clear sign of substantial freezing, and a warning to polar mariners that winter is certainly on the way.

If temperatures stay low for a few days, the pancakes meld and thicken to form floes of new ice. In a month, the ice may thaw, freeze, and be covered by snow, forming a sheet of sea ice 15–60 centimeters (6–25 in) thick. Sea ice that is attached to the edge of the continent is called fast ice; ice that drifts with the currents offshore is called pack ice. Fast ice thickens mostly as the result of new ice forming on the underside of the surface ice (this is also particularly true of the way Arctic sea ice forms), whereas 50 percent or more of the thickening of pack ice is due to the pancakes or small floes piling on top of one another (rafting) during storms.

This expanse of new ice grows at an incredible rate: up to 60 square kilometers (23 sq. miles) per minute. Throughout the winter, it advances about 4 kilometers (2½ miles) each day, adding almost 100,000 square kilometers (40,000 sq. miles) of new ice. In October, at the end of an average winter, it has effectively more than doubled the size of Antarctica. The 14 million square kilometers (5,400,000 sq. miles) of land abut 20 million

▼ STRONG ICE

By midsummer, there will be open water around the Cape Royds coastline (with Orcas and minke whales feeding) but in spring the fast ice is strong enough to travel safely on. Broken floes are here trapped in the sea ice, whose uneven surface and cracks bear testimony to some of the pressures acting on it.

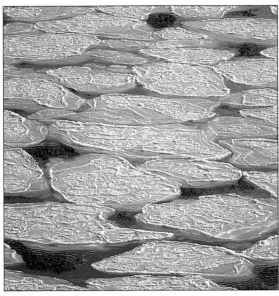

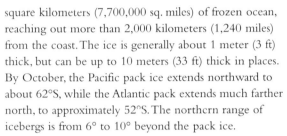

◄ PANCAKE ICE
PANCAKE ICE
Pancake ice looks very much like pale waterlily leaves—each flat plate has a raised edge around it from collisions with its neighbors.

Far left

◄ NEW ICE FOR OLD
Beyond the edge of last winter's sea ice in McMurdo Sound, new sea ice is forming. The appearance of new sea ice depends on conditions at the time. On calm seas a thin slick forms, which gradually thickens into glossy sheets just a few centimeters thick. Typically, disturbances from sea or wind break these fragile sheets up and they ride over each other, creating amazing patterns.

square kilometers (7,700,000 sq. miles) of frozen ocean, reaching out more than 2,000 kilometers (1,240 miles) from the coast. The ice is generally about 1 meter (3 ft) thick, but can be up to 10 meters (33 ft) thick in places. By October, the Pacific pack ice extends northward to about 62°S, while the Atlantic pack extends much farther north, to approximately 52°S. The northern range of icebergs is from 6° to 10° beyond the pack ice.

New ice does not accumulate uniformly around the whole continent. It starts to form in the Weddell Sea, and then in the Ross Sea and the Bellingshausen Sea. As Ernest Shackleton had ample time to observe, sea ice moves continually. In the Weddell Sea, where his *Endurance* was trapped and finally crushed, a clockwise drift is more defined than the drift in either the Ross Sea or the Bellingshausen Sea.

▲ WIDE WATERS
Even at the coldest time of
year there is a polynya (a
large area of open water) at
McMurdo Sound and Ross
Island. The nearest part of
Ross Island is Cape Bird, the
site of large Adélie penguin
colonies in summer. The high
points of Ross Island are
Mount Terror (center) and
Mount Erebus (right).

Nor is the spring melt a simple reversal of the freezing process. Ice that once formed near the coast can be found at huge distances from the shore by December, and in late summer, there can be a thick ring of pack ice far from land, while the coastal waters are relatively ice-free and navigable.

Polar dissimilarities

Sea ice of the two polar regions differs in many ways. Arctic sea ice may have a life span of up to eight years, but most Antarctic sea ice is less than a year old—only within the Weddell and Bellingshausen seas is sea ice found that is up to three years old. About 90 percent of Arctic sea ice is older than a year, more than 2 meters (6½ ft) thick, very strong, and low in salinity. A similar proportion of Antarctic sea ice is less than a year old, less than 2 meters (6½ ft) thick, structurally weak, and quite saline. Old sea ice is much stronger than first-year ice, and is much more difficult for ships to break through.

In winter, the area of Antarctic pack ice is much larger: 20 million square kilometers (7,700,000 sq. miles), compared with 14 million square kilometers (5,400,000 sq. miles) in the Arctic. In summer, the sea ice in both regions is much reduced: 4 million square kilometers

(1,500,000 sq. miles) in Antarctica, as against 7 million square kilometers (2,700,000 sq. miles) in the Arctic.

Sea ice nomenclature

Sea ice is named by size. Fragments of ice less than 2 meters (6½ ft) across—smaller than a grand piano—are called "brash ice." Ice floes are "small" if they are less than 100 meters (330 ft) wide, "medium" if less than 300 meters (990 ft) wide, and "large" if less than 2 kilometers (1¼ miles) wide. A "vast floe" measures up to 10 kilometers (6¼ miles) across; anything larger than that achieves "giant floe" status.

When wind and ocean currents push the ice apart, the open water is called a lead, and the heat released from the water below can create frost smoke. A really large area of open water is known by the Russian name: *polynya*. In shipping terms, "open pack" is water containing up to six-tenths ice, and "close pack" is seven- or eight-tenths ice, when some water is still visible. An icebreaker or ice-strengthened vessel is needed for movement in "very close pack"—nine-tenths ice. "Compact" pack is all ice, no leads; no water at all is visible—even most icebreakers would have difficulty passing through it.

Winter–summer sea ice

The continent always retains some sea ice in the sheltered areas around its coast, even in summer. In winter, when the sea freezes, the effective size of the Antarctic Continent increases dramatically, and the continent itself becomes locked away, largely unreachable by either human or animal life, for several months.

How sea ice behaves

In some ways, it is miraculous that sea ice forms at all. Only a few substances are less dense as solids than they are as liquids; water is one of them. The density of pure ice is 91.7 percent that of pure water—so it floats with 8.3 percent of its mass above the water. Further, because of its salt content, seawater freezes at about −1.8°C (29°F) and, unlike fresh water, increases in density as it approaches freezing point. So when it cools, it sinks. Only when the cold air above has cooled a top layer of water around 3 meters (10 ft) thick to freezing point can sea ice form.

The dynamics beneath sea ice are very different from those in a lake or river covered in ice. As fresh water cools, it increases in density until about 4°C (39° F), then it slightly decreases in density until 0°C (32°F). The density of salt water increases between 0°C and its sub-zero freezing point. So, in an ice-covered lake the coldest water is lighter than the warmer water, and the water forms bands: ice on top, colder water below it, and warmer water on the bottom. In the sea, the ice cools the water directly below it, some of which freezes to the bottom of the ice, while the rest falls (when its density increases as it approaches freezing point), to be replaced by warmer, lighter water. So the water is constantly circulating below the sea ice.

The extent of sea ice varies considerably from year to year, and climatologists are looking for patterns that may reveal evidence of global warming. The presence of so much ice certainly influences the climate of Antarctica. The sea ice "skin" restricts the normal heat exchange between the ocean and the atmosphere. In spring, the sea ice reflects back solar energy that would otherwise warm the ocean and start summer earlier. At the start of winter, new sea ice reflects the last sunlight away, and isolates the (relatively) warm ocean from the air, so winter comes much sooner.

Is sea ice salty?

Sea ice always contains some salt, but its salinity is generally about one-tenth that of seawater. As seawater freezes, the salt is forced out but gets caught up in tiny brine pockets within the ice. The faster the freezing, the more salt is captured. Over weeks and months, the trapped salt migrates downward through the ice in response to gravity. So, gradually, the sea ice becomes less saline at the top than at the bottom. In the Arctic, the Inuit often use surface ice for their drinking water; sea ice that is more than a year old has a salt content of less than 0.1 percent.

Freezing and melting of polar sea ice greatly affect surface-water salinities of the sea. As sea ice forms, the exclusion of salts from the growing crystals of ice concentrates the salts into the seawater, increasing the sea's salinity. Through spring and summer the surface seawater loses salinity as this sea ice melts.   DM

▼ BEWARE OF CRACKS!
A recurring crack in Erebus Bay shows that even the fast ice is subject to pressures such as tides, currents, and weather. Fast ice is a convenient traveling surface between winter and early summer—but care is necessary.

▲ SHATTERED ICE
Angular in shape, the ice forms collide with one another. These floes may freeze together and break up several times before forming a solid cover. As the ice thickens, ocean swells break it into larger pieces.

Summer minimum sea ice
Winter maximum sea ice

Haakon VII Sea

Weddell Sea

Ronne Ice Shelf

Bellingshausen Sea

Amery Ice Shelf

Davis Sea

Shackleton Ice Shelf

Amundsen Sea

Ross Ice Shelf

Antarctic Circle

Ross Sea

Dumont d'Urville Sea

0°
30°W
30°E
60°W
60°E
90°W
90°E
120°W
120°E
150°W
150°E
180°
70°S
80°S
90°S

Global warming

Today, the earth supports six times more people than it did 200 years ago. Between 1760 and 1820, the population of Britain jumped from 6 million to 14 million, its agrarian, handicraft economy becoming one dominated by industry and machines. The Industrial Revolution rapidly spread across the world, increasing the demand for energy. The burning of wood, coal, oil, and natural gas generated carbon dioxide.

Greenhouse gases

Carbon dioxide is one of the greenhouse gases that occur naturally in the atmosphere, absorbing heat released by the land and oceans in a process known as the greenhouse effect. Without the heat-trapping properties of these gases, the earth would be too cold for human habitation.

However, atmospheric carbon dioxide concentrations are now more than 30 per cent higher than they were 200 years ago. A typical car, for example, releases tonnes of carbon dioxide each year. Methane concentrations are 140 percent greater; cows burp as much as 100 kilograms (220 lb) of methane per year. New greenhouse gases, such as chlorofluorocarbons (CFCs), have been added to the air. This extra greenhouse effect is warming the lower atmosphere and changing earth's climate.

Likely climatic changes

Average surface temperature has already risen by between 0.3°C (0.5°F) and 0.6°C (1°F) during the past 100 years; the twentieth century was the warmest of the past millennium. In the northern hemisphere, the 1990s was the warmest decade, and 1998 was the warmest year. World rainfall patterns are changing, sea level is rising, and glaciers are retreating.

By the end of the twenty-first century, global temperatures are likely to be between 1.4°C (2.5°F) and 5.8°C (10°F) higher than now. This anticipated warming

It is anticipated that global temperatures will rise significantly by the end of the twenty-first century. Land will warm more than the sea, and the greatest warming is likely to occur near the Poles. Luxuriant vegetation, such as the above, is now found on only a few sub-Antarctic islands but may well become more widespread.

will be greater than any recent natural fluctuations, and will happen faster than any other known changes experienced since the last Ice Age.

Land will warm more than the sea, with the greatest warming expected to occur in regions close to the Poles. The temperature difference between night and day is likely to decrease. Arid and semi-arid parts of Africa, southern Europe, and the Middle East, and areas of Latin America and Australia will become drier as evaporation increases.

A warmer atmosphere will heat the oceans' upper layers, and the expanding water will raise sea level. Land-based ice in temperate regions such as South and North America and Greenland will melt more rapidly, adding to this rise. Conversely, increased precipitation over Antarctica and Greenland will lock water away in the ice sheets. Scientists estimate that by 2100, sea level will be 9–88 centimeters (3–35 in) higher.

Almost everywhere, these weather and climate differences will require people to adapt significantly. The changes will be even more serious for the natural environment, as climate change will exacerbate pressures on ecosystems from population growth and human exploitation. PH

When the ice melts

The huge weight of the polar ice sheets depresses the earth's crust—remove the ice and the crust will rise again to its original level. The shoreline terraces of eastern Canada's Great Lakes formed by this process of "glacial rebound." They tilt gradually northward from near Chicago to a maximum of about 300 meters (990 ft) near Hudson Bay, and reflect the disappearance of the 10,000-year-old continental ice sheets of North America. This model indicates that the rock surface of Antarctica would rise by glacial rebound when the ice melts. The uplifted crust would displace seawater and the sea level would rise, but in small amounts and extremely slowly. Under the thickest ice of East Antarctica the crust is estimated to be depressed as much as 950 meters (3,100 ft). How much sea levels would rise from that amount of uplift is unknown.

Melting ice would return water formerly held on land to the world's oceans and, again, the sea levels would rise. Doubtless, this would begin slowly, as climates warmed sufficiently to begin melting the polar ice sheets. Warming and thermal expansion of the world's oceans might alone result in a rise of some 0.5–4 meters (1½–13 ft), but it might take a thousand years for atmospheric warming to reach oceanic bottom waters. Already, however, sea levels are rising almost imperceptibly. A quickening rate would have serious consequences for the world's coastal cities.

Antarctica contains 90 percent of the world's ice, enough to affect sea levels seriously. Estimates of amounts and rates vary. Calculations suggest that sea levels rose about 130 meters (425 ft) from the melting of 21,000-year-old, northern hemisphere ice of the last Ice Age. The melting of all today's Antarctic ice might raise the sea level by about 80 meters (260 ft).

The East Antarctic ice sheet lies on bedrock that is mostly at sea level or above, whereas much of West Antarctica's ice is based on bedrock well below sea level, and in places far below it. The decay of the ice in East Antarctica might result in a 60-meter (200-ft) sea level rise, but at a very slow rate, perhaps over 10,000 years, as the ice would remain mostly grounded. Glaciologists worry, however, that West Antarctica's ice could melt at a catastrophic speed due to flotation from its bedrock below sea level. A rise in sea level of approximately 15 meters (50 ft) might occur in a few tens to a hundred years. The melting of floating ice, such as the immense Ross Ice Shelf, would not affect sea levels, as the ice is already in flotational equilibrium.

When its ice does melt, Antarctica will be a much smaller continent than it is today. East Antarctica will be the main landmass, but joined to a peninsula of today's Ellsworth Mountains. Today's Antarctic Peninsula will be a separate island mass, and the Marie Byrd Land region will be a sea dotted with volcanic islands, separated from the "Antarctic Peninsula" by a deep marine trough, even after glacial rebound. Many mountain ranges now under ice, such as the Gamburtsev Subglacial Mountains, would rise above the plains of this continent. **AF**

▼ **AN AWFUL LOT OF ICE**
Ninety percent of the whole world's ice can be found in the Antarctic. Scientists estimate that, if all the ice in Antarctica were to melt, sea levels could rise by as much as 80 meters (260 ft).

The Antarctic ozone hole

Ozone, a form of oxygen, is a comparatively sparse constituent of the earth's atmosphere—there are only about three molecules of it in every ten million molecules of air. However, ozone is extremely important to the atmosphere and, for the past 600 million years or so, has been the atmospheric sentinel that has protected life on the planet from the biological damage caused by the sun's ultraviolet radiation. Many experimental studies of plants, animals, and humans have shown that there are harmful effects from excessive exposure to ultraviolet-B radiation. This is why the maintenance of the ozone layer is of vital concern, particularly in view of recent evidence that synthetic chemicals have caused serious loss of ozone from the atmosphere.

Damaging the atmosphere

During the early 1970s, atmospheric scientists engaged in theoretical research first became aware that possible damage to the ozone layer might be occurring as a result of human activities. However, it was 1985 before the British Antarctic Survey (BAS) discovered the Antarctic ozone hole and solid evidence was found that the ozone layer might be under threat. In 1987, the United States carried out a research campaign, where NASA aircraft, fitted with sophisticated scientific equipment, flew through the stratosphere from South America into the ozone hole. These experiments produced irrefutable evidence that the ozone hole was mainly caused by the chemical destruction of ozone by atmospheric chlorine. Because the quantity of chlorine required to destroy this amount of ozone could not originate from natural sources, another culprit had to be identified, and scientists then reached the inescapable conclusion that pollution from synthetic chemicals was causing ozone depletion. The prime offenders were the so-called halocarbon compounds, such as the chlorofluorocarbons (CFCs) that contain chlorine atoms.

CFCs were nontoxic and nonflammable and could easily be converted from a liquid to a gas and vice versa. They had many useful industrial and commercial applications; for example, as foam-blowing and cleaning agents, as the cooling fluid in refrigerators, or as aerosol-spray propellants. Because CFCs were designed to be relatively chemically nonreactive with most substances in the environment, this also meant that any of these substances would last a very long time in the atmosphere if they escaped or were released.

In the atmosphere, natural atmospheric circulation transports these chemicals high into the stratosphere, where the ultraviolet radiation is comparatively strong and can strip chlorine atoms off the CFC molecules. Thus, chlorine atoms from synthetic chemicals are liberated into the stratosphere, and are able to attack ozone molecules immediately, and convert them to oxygen by pulling off single oxygen atoms. Through further chemical reactions in the atmosphere, these chlorine atoms are effectively recycled to destroy more ozone molecules— one free chlorine atom in the stratosphere can destroy tens of thousands of ozone molecules before nitrogen compounds remove it from its destructive cycle. This catalytic process is the dominant, fundamental mechanism of ozone destruction. Through it, large numbers of ozone molecules are converted into oxygen, and during the conversion their ability to absorb ultraviolet radiation is lost. Ozone depletion is also caused, to a lesser extent, by pollution from chemical components other than chlorine, such as bromine. As a result, ultra-violet radiation at the earth's surface increases to a level that has the potential to be detrimental to life.

▲ TELLING INSTRUMENT
A scientist measures how much ozone is overhead at the Arrival Heights Laboratory in Antarctica, using a Dobson spectrophotometer. The Antarctic ozone hole was first discovered from Dobson spectrophotometer data.

Making ozone

The sun's ultraviolet radiation is more intense in the upper atmosphere, where it is not absorbed or reflected by atmospheric gases and particles to the extent that occurs lower down. This means that it can separate the atoms in oxygen molecules more often than it can lower down in the atmosphere, and the result is more free oxygen atoms. However, the joining of free oxygen atoms with oxygen molecules to produce ozone occurs more often lower down, where the atmosphere is denser, and these atoms and molecules collide more often. Because there are more free oxygen atoms higher up, yet more ozone-producing collisions between oxygen atoms and molecules lower down, the creation of ozone is at its greatest at the in-between altitudes—and it is this that forms an "ozone layer." The altitude range of the atmosphere in which the ozone layer resides is between about 10 and 50 kilometers (6 and 31 miles), and is known as the "stratosphere."

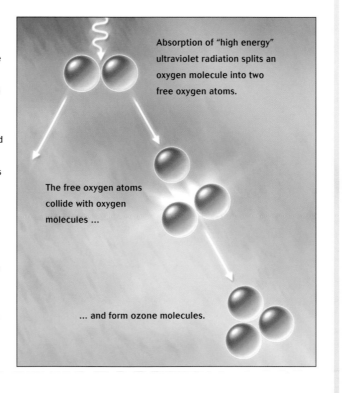

Absorption of "high energy" ultraviolet radiation splits an oxygen molecule into two free oxygen atoms.

The free oxygen atoms collide with oxygen molecules ...

... and form ozone molecules.

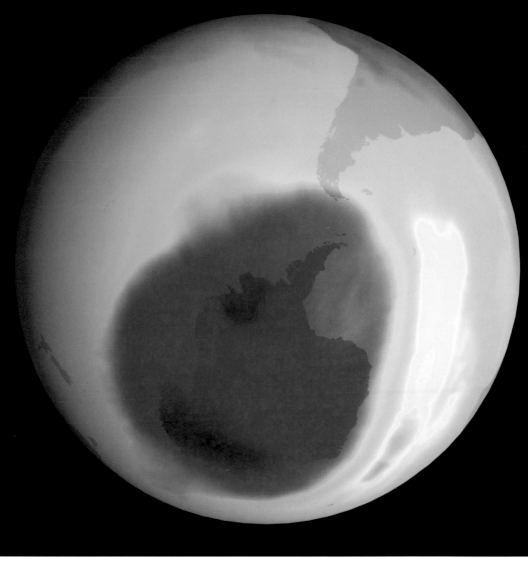

> NASA RECORDS LARGEST OZONE HOLE TO DATE
In early September 2000 the Antarctic ozone hole reached a record size of about 29 million square kilometers (11 million sq. miles)—more than three times larger than the land area of the United States. The ozone hole (dark blue) can be seen over the southern tip of Chile, and causes ultraviolet radiation increases.

Patterns of distribution

Ozone depletion in the stratosphere depends on latitude and season. For example, at latitudes 30°S to 60°S, total ozone reductions year-round were about 5 percent between the late 1970s and the mid-1990s.

The Antarctic ozone hole forms in early spring over the Antarctic Continent when the polar dawn occurs in August and sunlight splits chlorine molecules into the ozone-destroying chlorine atoms that drastically deplete ozone in the stratosphere. In the years leading to and including 2000, the ozone hole has suffered an ozone loss of up to 70 percent, and has sometimes covered 29 million square kilometers (11,300,000 sq. miles), more than three times the area of the United States. The ozone hole slowly reaches its greatest depth and area in late September to early October, and then slowly recovers as fresh ozone from lower latitudes gradually enters the ozone hole region. On occasions, the ozone hole passes over Chile in South America, where people are alerted through the media to the concern of increased ultraviolet radiation. Typically, the ozone hole finally breaks up after mid-November, when the wind system that contains it (called the polar vortex) weakens and breaks down. It is at this time that ozone-depleted air can move to lower latitudes and reduce ozone levels over Australia, New Zealand, and other southern hemisphere countries. The effect of this ozone decrease appears to persist into the following year.

International problem solving

The evidence that synthetic chemicals are responsible for the ozone depletion observed since the mid-1970s is now overwhelming, and an international treaty was proposed to progressively eliminate all known ozone-depleting substances. Parties to the Vienna Convention for the Protection of the Ozone Layer (1985) and the Montreal Protocol on Substances that Deplete the Ozone Layer (1987) now number 155, of which more than 100 are developing countries. It is the first time that nations have agreed, in principle, to cooperate to solve a global environmental problem.

Evidence is now mounting that the so-called greenhouse effect may be reducing the recovery of global ozone by cooling the stratosphere where the ozone layer resides, and making the chemical reactions that cause ozone depletion more efficient. The greenhouse effect may also result in more water entering the stratosphere over the tropics, further exacerbating the chemical processes that destroy ozone.

These effects, coupled with the longevity of many of the ozone-depleting substances in the earth's atmosphere and the burden of future emission of chemicals (even though measures are now in hand to reduce such emissions), mean that the length of time required for stratospheric ozone to recover to historical levels (pre-1980) will depend on many factors. Current computer model estimates place such a recovery around the middle of the twenty-first century, but many discoveries were made during the 1990s that significantly altered the experts' understanding and forecasting of ozone depletion, and more of these may be yet to come.

Whatever the future outcomes of research, it is vital that people cease to release CFCs and other ozone-destroying chemicals into the stratosphere so that the ozone layer has a chance to repair itself. PL

▲ EXTRA LOAD
Daily measurements of Antarctic ozone up to 35 kilometers (22 miles) are made using a miniature chemical processing package called an ozone sonde. These can be suspended from conventional weather balloons, as shown here by members of the University of Wyoming research team.

Lights in the sky

Aurora, named for the Roman goddess of dawn, is a dramatic natural phenomenon that swirls or pulsates in the night sky, creating curtains and patches of colored light, predominantly in mixtures of green, red, and violet. An active display is an awe-inspiring sight.

More analytically, aurora is the light emitted by atoms, molecules, and ions that have been excited by charged particles, principally electrons, traveling along magnetic field lines into the earth's upper atmosphere. Light is emitted when excited atmospheric constituents release energy while returning to a lower energy state. This is similar to the process that generates light in a neon tube. For neon, the dominant wavelength emitted corresponds to a red color

▼ CELESTIAL CURTAINS
Reflected in the newly frozen bay, sweeping arcs wave majestically as the expansive phase of an auroral substorm begins. The individual bands brighten and develop billowing folds, often twisting into spectacular spiral formations as waves of light surge along their entire length.

What causes aurora?
The solar corona—the outer region of the sun—is hot enough to continually emit solar wind plasma (a gas consisting of charged particles composed mainly of protons and electrons), which streams into space at speeds of around 300 kilometers (187 miles) per second. The earth's magnetic field also confines plasma, which travels along the field lines and is bounced between the hemispheres by the increasing field strength encountered in polar regions.

Electrons in the solar wind or trapped by the earth's magnetic field do not typically have sufficient energy to generate aurora. Processes associated with the interaction of solar wind and earth's magnetic field accelerate these-

electrons. Solar plasma also carries a magnetic field. When the solar wind's magnetic field opposes the earth's magnetic field, energy is transferred by a process called magnetic reconnection. The extraction of energy from reconnecting magnetic fields is readily demonstrated by gradually bringing the opposite poles of two bar magnets together. Earth's magnetic field becomes linked to the solar wind by magnetic reconnection on the day-side and disconnected on the night-side, accelerating electrons on both occasions. Electrons may also be accelerated by an interaction that resembles friction, as solar wind plasma that has not been magnetically reconnected to the earth races past the flanks of its magnetic field. This manner of energy transfer may be likened to wind passing over water and generating waves.

Global occurrences

Aurora, in the form of an oval around the magnetic pole in both hemispheres, results from these interactions. Most commonly, aurora is located at magnetic latitudes of about 67° on the night-side and about 75° on the day-side. When considering the chances of observing aurora at a particular location, it is important to account for the separation of magnetic and geographic poles.

Aurora is known as aurora australis, or the southern lights, in the southern hemisphere, and as aurora borealis, or the northern lights, in the northern hemisphere. Auroral activity in the two hemispheres is strongly linked in space and time, as electrons can travel freely in either direction along magnetic field lines. Brightenings and movements in the two hemispheres coincide within seconds in magnetically linked regions. This was confirmed in the 1970s by measurements made on specially equipped aircraft following flight paths linked by the earth's magnetic field out of Christchurch, New Zealand, and Fairbanks, Alaska.

Sunspots (cooler regions of intense magnetic field) are signs of enhanced solar wind activity. They have a cycle of about 11 years, which most recently peaked in 2000. Aurora is more common at lower latitudes at peak sunspot times and during the subsequent two years; it slightly intensifies in spring and autumn.

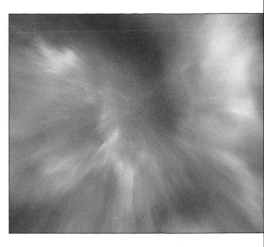

Observing aurora

From the ground, it is possible to observe about 700 kilometers (435 miles) of the approximately 8,000-kilometer (4,970-mile) wide auroral oval. In quiet times this appears as a shimmering curtain extending in an east–west direction.

More active displays, known as auroral substorms, occur when stored energy is rapidly released. In an explosive onset, rays and arcs may twist, move, and break into small segments as enhanced electric currents flowing near auroral heights modify the field lines along which charged particles are traveling. Pulsating patches of auroral glow may persist during the gradual recovery phase, and eventually quiet arcs re-form. A substorm may last an hour, and may recur up to three or four times each night.

The aurora produces many colors, each corresponding to specific energy level transitions of excited atoms, molecules, or ions. Three, however, are most significant. A green line emitted by an oxygen atom is dominant when a sharp lower border can be discerned, usually at about 100 kilometers (60 miles). Red is generated by another atomic oxygen transition, and is prominent at altitudes above 250 kilometers (155 miles). A violet band emitted by a nitrogen molecular ion is about five times less intense than the green line, but can be observed on the leading edge of active aurora because of the delay of about a second in the average emission time of auroral green light. GB

> **A GIFTED DOCTOR**
Edward Wilson, a close confidant of Robert Falcon Scott, was a medical doctor, a naturalist, and an artist. During Scott's expedition of 1901–04, Wilson sketched the *Discovery* beneath a stunning aurora australis.

⌃ **CRIMSON CORONA**
The expansive phase of an auroral substorm develops in minutes to the break-up stage. Individual folds in previously quiescent auroral arcs can split into spirals that splay outward from a single point.

Above left

⌐ **VANISHING POINT**
The coronal point is where overhead auroral rays seem to meet. Along the line of the earth's local magnetic field in this direction, individual rays delineate neighboring field lines that appear to converge like skyscraper spires viewed from immediately below.

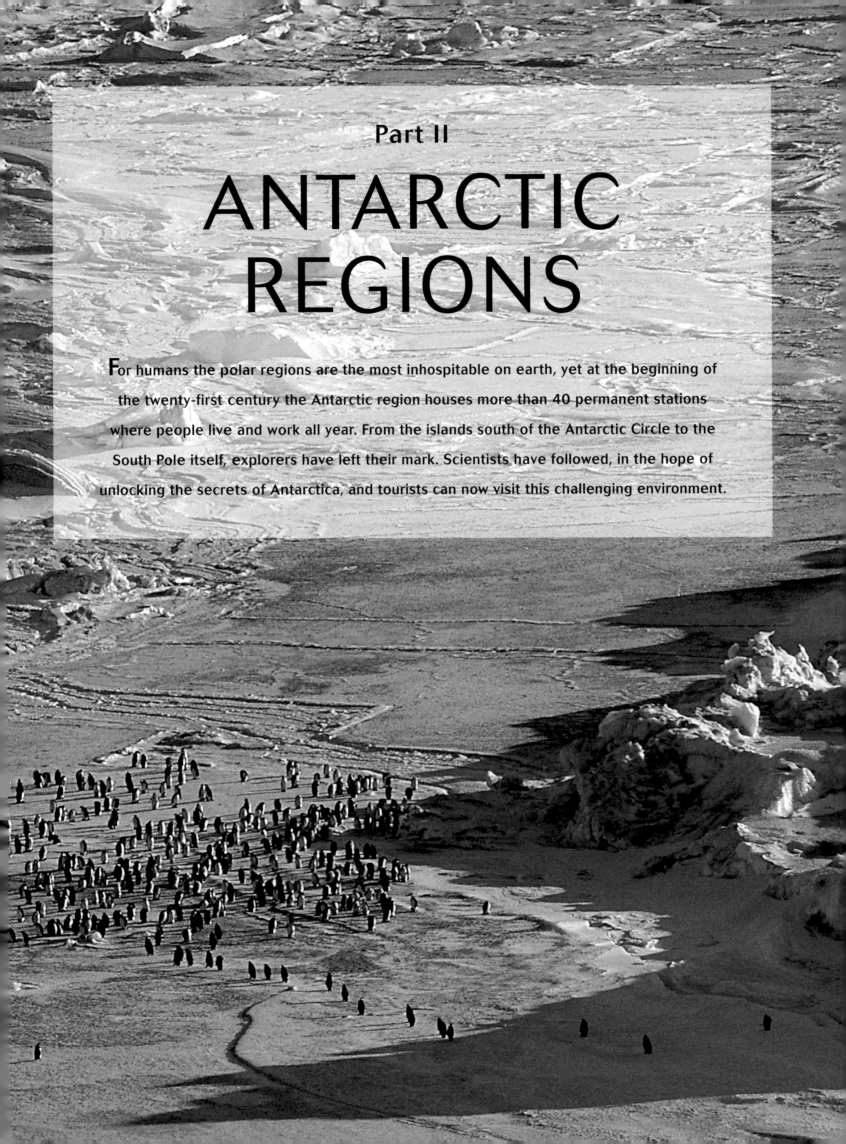

Part II
ANTARCTIC REGIONS

For humans the polar regions are the most inhospitable on earth, yet at the beginning of the twenty-first century the Antarctic region houses more than 40 permanent stations where people live and work all year. From the islands south of the Antarctic Circle to the South Pole itself, explorers have left their mark. Scientists have followed, in the hope of unlocking the secrets of Antarctica, and tourists can now visit this challenging environment.

The Antarctic Peninsula

Previous pages

◄ **PENGUINS AND PEOPLE**
Cape Royds, the westernmost
point of Ross Island, is home
to the world's southernmost
Adélie penguin rookery (about
4,000 breeding pairs). Not far
from the colony is Ernest
Shackleton's hut, which was
abandoned in 1909 after
Shackleton almost reached
the South Pole.

The north

The notch in the tip of the Antarctic Peninsula is Hope
Bay, discovered in 1902 by the Swedish expedition led
by Otto Nordenskjöld. He named it Hope Bay, after
three of his party inadvertently wintered there during
1903; they were dropped off not long before their ship,
Antarctic, sank and the sound beyond the bay bears the
name of their doomed vessel. Their makeshift stone hut
stands near the dock—it was largely rebuilt over the
summer season of 1966–67.

Situated in the Antarctic Sound, some 15 kilometers
(9 miles) southeast of Hope Bay, Brown Bluff is where
the eastern edge of the Tabarin Peninsula drops almost
sheer to the water from an ice-capped summit. From a
rocky beach, a steep scree slope rises to a towering, rust-
colored cliff of volcanic rock. This could be central
Australia or the United States Badlands, were it not for
the penguins and Weddell seals along the shore and the
ice cap high above.

Adélie penguins number tens of thousands on Brown
Bluff, and several hundred Gentoo penguins also live
there. The slope is covered with loose rubble and rock

slips are common. Scientists believe landslides and falling
boulders may have obliterated some penguin rookeries.

James Clark Ross, aboard *Erebus* in search of the
South Magnetic Pole, named Paulet Island after Lord
George Paulet of the British Royal Navy. A circular volcanic cone, the island rises from the sea 3 kilometers
wide and 353 meters high (1¾ miles and 1,158 ft).

This is as far as many Antarctic travelers will be able
to venture into the Weddell Sea, because the clockwise
current ensures that even this northern region has its
share of spectacularly large icebergs. Huge tabular bergs
encountered on the way through the Antarctic Sound
give some idea of the difficulties faced by the early
explorers.

One of Antarctica's largest colonies of Adélie
penguins occupies Paulet Island. At least 100,000 breeding pairs crowd the beach and nearby steep slopes for
the short summer—a fine example of avian tenement
living. Snowy sheathbills, skuas, and Giant petrels live off
the rookery, and opportunistic Leopard seals patrol the
shores in the certainty of easy prey.

French explorer Dumont d'Urville sighted Astrolabe
Island in the Bransfield Strait during his 1837–40
Antarctic voyage, and named it after his expedition flagship, *Astrolabe*. The rocky island has cliffs plunging more
than 100 meters (330 ft) directly to the sea. Chinstrap
penguins, fur seals, and Weddell seals breed on the low
ground, and skuas and fulmars nest along the craggy
heights. Shore landings are possible here.

Mikkelsen Harbor, on the south side of Trinity
Island, was named by the 1901–04 Swedish South Polar
Expedition led by Otto Nordenskjöld. The term harbor
is deceptive, as the landing site is in fact a small island in
a bay that has no harbor at all. Rocky shoals lead to
a rocky beach, with glaciers forming a magnificent
backdrop on a fine day.

There are a few unoccupied huts on
the island, and a radio mast at its low
summit. The harbor's chief residents are
Gentoo penguins on its slopes, and
slumbering Weddell and elephant seals
on the beach.

◄ **CHILLING OUT**
Out of the water, Weddell
seals, like this dozing
pup, appear to do little
more than rest and sleep.
They have been observed
on the beach below
Brown Bluff, and are
common in most locations
along the northeastern
end of the Antarctic
Peninsula.

◄ **BUILDING BLOCKS**
When *Antarctic* sank in 1903, Captain
Carl Larsen and his crew of 20 sailors
(and a cat) lived in this small hut on
Paulet Island for nine months
over a long polar winter. Fortunately for the stranded men,
building the hut was quite easy:
regular freezing and thawing
of the island's basalt
caused the rock to break
into even-sized flat stones—
ready-made building blocks.

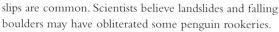

66°S

72°W

68°S

72°W

ADELIE PENGUINS AT PLAY
Hope Bay is one of the few places on the mainland of the Antarctic Peninsula where large numbers of penguins can be seen.

Antarctica Peninsula

The Gerlache Strait

Curtiss Bay, at the top of Gerlache Strait, is often crowded with large icebergs; whales visit regularly, and Leopard and other seals rest on the floes. The bay is little visited, but to the north there is a small rocky promontory covered in snow where landings are possible.

Crammed with spectacular icebergs, Cierva Cove lies at the top of Hughes Bay on the southern side of Cape Herschel. Cierva Cove is the site of Argentina's summer station Primavera.

In 1960 the British named the low-lying, bare Hydrurga Rocks, very near the much larger Two Hummock Island, after the Leopard seal: *Hydrurga leptonyx*. Leopard seals are present off-shore; they prey on the Chinstrap penguins that establish small rookeries wherever bare rock is accessible, as it is here. Antarctic shags also nest on Hydrurga Rocks.

On the western coast of Graham Land on the Reclus Peninsula, Portal Point, the entrance to Charlotte Bay, received its name from the British after they built a tiny refuge hut there in 1956 at the start of a route they established up to the plateau. A survey group wintered there in 1957, but now only the foundations of their hut remain. Mountains, glaciers, and icebergs make Charlotte Bay one of the most attractive inlets along the Antarctic Peninsula.

South of the Reclus Peninsula, Wilhelmina Bay is frequented by whales and littered with islands that provide some shelter in almost any weather. The only exception is when katabatic winds howl down from the Forbidden Plateau, more than 2,100 meters (7,000 ft) above the bay, and whip the water into foam.

Orne Harbor, discovered by Adrien de Gerlache in 1898, is a mere indentation on the west coast of Graham Land. Although there are no landing sites except near Spigot Peak, and not much wildlife, it offers some excellent views of glaciers from cruising Zodiac craft.

Narrow Errera Channel contains Cuverville and Danco islands, and is one of the Antarctic Peninsula's most frequented sites by visitors longing to see pods of Humpback whales. Errera Channel is an awesome environment, with ice in the water and towering mountains.

65°S

◄ **BENEATH THE ERRERA ICE**
In temperate waters, diving time is governed by air and depth, but in the near-freezing waters of Antarctica, the condition of a diver's hands usually limits diving time to no more than 20 minutes—after that, pain and numbness set in.

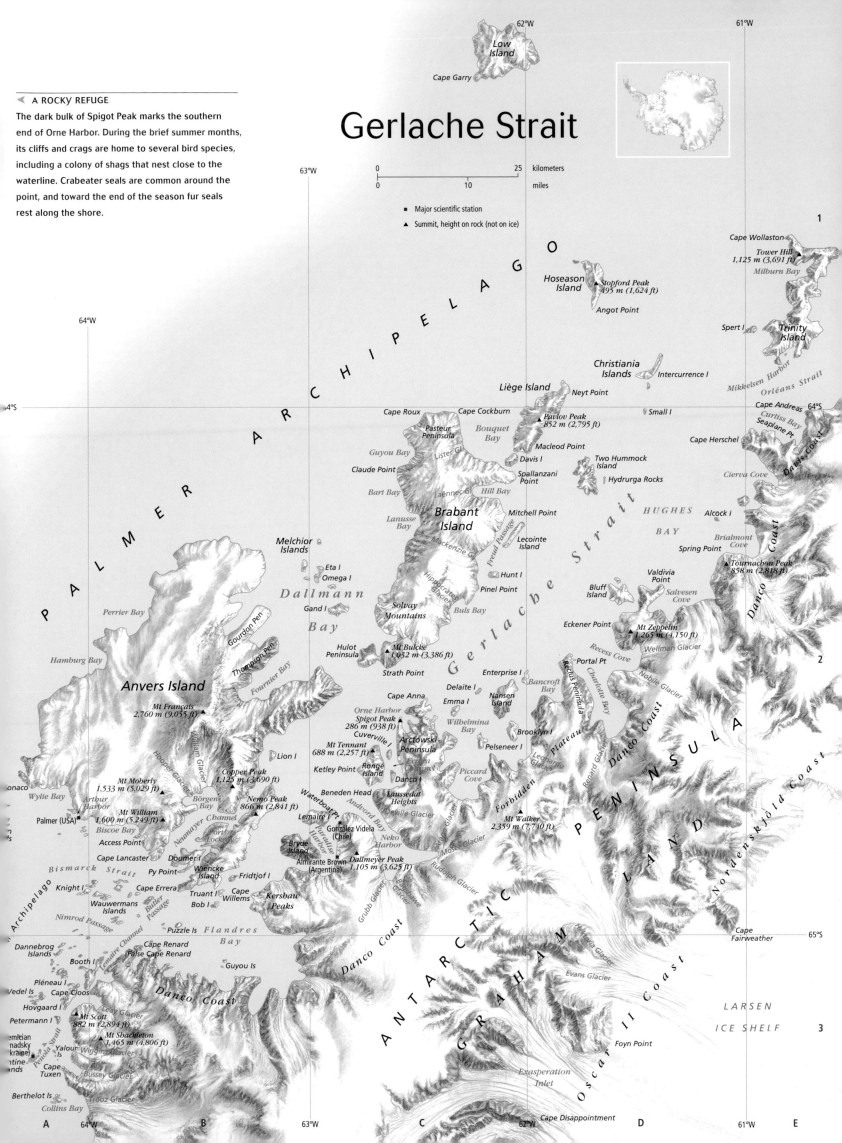

Gerlache Strait

Cuverville Island stands like a stopper in the northern end of the Errera Channel. Five thousand pairs of Gentoo penguins breed on the long, shingly beach, and in coves at the northern end of the island. In late summer, when most of the snow has melted, patches of vegetation are visible. Mosses and lichens are joined by Antarctica's two flowering plants: Antarctic hair grass (*Deschampsia antarctica*), and the pearlwort (*Colobanthus quitensis*). Like all Antarctic vegetation, these plants are extremely slow-growing, and may take years to recover from the crushing damage of a single footstep.

Danco Island was charted by de Gerlache's *Belgica* expedition in 1898, and later named by Britain for Lieutenant Emile Danco, the Belgian geologist who died on board *Belgica* while she was trapped in the ice of the Bellingshausen Sea. Danco Island is closed to tourists.

Heading south from Errera Channel, one meets a deep expanse of water surrounded by mountains and glaciers spilling down to the shoreline. The northern shore of Andvord Bay provides some of the most awe-inspiring scenery in the whole Peninsula. Distances are deceptive, and it is hard to grasp the sheer scale of this astonishing landscape. The roar of an avalanche or the crash of a calving iceberg may be audible well after the event—proof of the distance the sound has traveled, although the scene may appear closer.

Tucked within the eastern side of Andvord Bay is the sheltered cove of Neko Harbor, a place of great beauty, which nevertheless carries a name associated with slaughter: *Neko* was a whaling factory ship that operated in this region between 1911 and 1924, and frequently moored in this protected bay. During those years the pristine landscape of today must have been a scene of death and decay, besmirched by stinking smoke and lapped by bloodstained water.

Coming from the north, the entry point to Paradise Harbor is Waterboat Point, now the site of the Chilean base of González Videla. The British Imperial Expedition of 1921–22 named Waterboat Point for the abandoned waterboat that housed two of them when the expedition collapsed. The few remnants of their boathouse remain near the station as a Historic Site. The bay is the true glory of this location—in still weather, its surface reflects peaks and glaciers, and the water is often crowded with icebergs, seals on floes, and even the odd whale.

Tough whalers were the first to name Paradise Harbor; it is also commonly—but incorrectly—referred to as Paradise Bay. Situated behind Bryde and Lemaire islands, the bay is indeed spectacular. Argentina's Almirante Brown base and Chile's González Videla at Waterboat Point together receive more tourists than anywhere else in Antarctica, partly because of the reputation of Paradise Harbor, but also because both places offer the chance to set foot on the Antarctic Continent.

Anvers Island and environs

It can be difficult at times to distinguish the mainland of the Antarctic Peninsula from the offshore islands of the Palmer Archipelago. In this world of black basalt and white glaciers, only charts may reveal the difference. This is true of Anvers Island, one of the largest islands north of the Antarctic Circle on the western side of the Peninsula. It receives few visitors except for those who have pre-arranged a visit to Palmer station, part of the United States Antarctic Program. The station has excellent marine tanks with krill, starfish, sponges, and sea spiders. From the deck near the aquarium, several seal species, including Crabeaters and Leopards, are usually visible; fur seals join them later in the season.

Lemaire and beyond

The tall, black twin peaks of Cape Renard mark the northern entrance to Lemaire Channel—one of the highlights of a voyage along the Antarctic Peninsula. Sometimes this narrow passage is impassable—currents and wind can fill the channel with sea ice and larger icebergs within hours—but when the ice permits, the hour's voyage each way is an unforgettable experience. The ice often supports basking Crabeater seals, and Orca, Minke, and Humpback whales navigate the passage.

The most spectacular feature of Pléneau Island is a field of grounded icebergs, west of Lemaire Channel. Because the Pléneau bergs are grounded, they are more stable than floating ones, and one can navigate Zodiac craft through them. They form a fairyland of arches and pinnacles and range in color from pellucid turquoise, where shallow water laps at their feet, to brilliant blue in the cracks in the ice. Pléneau Island is often the turning point for Antarctic cruises.

Antarctic Circle

The Antarctic Circle is only a theoretical line on a map, or a reading on a global positioning system (GPS), unless seen at the solstice, when the sun stays above, or below, the horizon. The polar circles move because the earth wobbles slightly on its axis. The Antarctic Circle is currently at 66°33′39″S, and is heading further south.   DM

▼ RESEARCH BASE
Palmer station is one of the few bases conducting long-term research close to the Antarctic Peninsula itself, rather than on King George Island in the South Shetlands. One current study is investigating how Antarctic marine life is being affected by changes in sea ice cover.

▶ PRIVATE WONDERLAND
A maze of islands, rocks, and grounded icebergs stretches south beyond Petermann Island. Accessible only to comparatively small, shallow-drafted yachts, the area has several sheltered anchorages with outstanding views.

TOWERING PEAKS

The spectacular northern entrance of the Lemaire Channel is marked by Cape Renard (in the middle of the photograph). Beyond the soaring twin peaks, the channel narrows dramatically to become "False Cape Renard. "

DISPUTED TERRITORY

The British survey hut at Port Lockroy was built in the middle of a Gentoo penguin colony, and the birds use the footpaths as their own. When staff arrive each spring, they must gently discourage a few penguin pairs from nesting on the wire-mesh front stairs.

GRISLY REMINDERS

A Gentoo penguin perches on a whale vertebra, one of thousands left strewn along the beaches of the Peninsula—and a reminder of days past, when quiet bays and inlets were used as gathering points for whaling operations.

The Ross Sea region

SLEEPING VOLCANO
A team of New Zealanders in a Hagglunds ATV (all-terrain vehicle) travel across the vast, flat Ross Ice Shelf toward Mount Discovery. This dormant volcano was named by Captain Scott after his ship—its youngest vents are 1.8 million years old.

The Ross Sea lies at the southernmost limit of the Pacific Ocean sector of Antarctica. When James Clark Ross sailed into the expanse of the sea that now bears his name, he hoped that it would allow him sailing access to the South Magnetic Pole. However, the impressive line of mountains that he saw to the west soon quelled this ambition. The western shore of the Ross Sea is delineated by the Victoria Land coast up to Cape Adare, and its eastern boundary is Cape Colbeck, at the northwest tip of Edward VII Land. Its total area is almost 1 million square kilometers (386,000 sq. miles). The Ross Sea is quite shallow—between 300 and 900 meters (1,000 and 3,000 ft)—but to the north it plunges sharply to depths of more than 4,000 meters (13,115 ft). At its southernmost edge is the Ross Ice Shelf, one of the most impressive features of Antarctica.

At the northwest corner of the Ross Ice Shelf is Ross Island, which is the most historically significant location in Antarctica. It was from here that Scott, then Shackleton, then Scott again led parties out on the first great attempts to walk to the South Geographic Pole—and it is here that Mawson, Edgeworth David, and Mackay returned after walking to the South Magnetic Pole. Today, it is the site of McMurdo Station, the largest Antarctic base. All-volcanic, the island is roughly the shape of a four-pointed star about 70 kilometers (43 miles) wide.

The sea ice around Ross Island does not break up until late January and re-forms in early April, and about six ships visit over the short summer season. Across its southern side the island is attached to the ice shelf from near the tip of Hut Point Peninsula to Cape Crozier.

Glaciers spill onto the ice shelf and onto the western shore. It is also linked by ice shelf to Victoria Land; the waterway in between known as McMurdo Sound is about 148 kilometers long and 74 kilometers across at its widest point (92 miles and 46 miles).

Mount Erebus looms over the island. Apsley Cherry-Garrard wrote after his first view of the mountain on Scott's last expedition: "I have seen Fuji, the most dainty and graceful of all mountains; and also Kinchenjunga; only Michael Angelo among men could have conceived such grandeur."

Hut Point Peninsula

Now a modern scientific station, Scott Base was built by Sir Edmund Hillary's Commonwealth Transantarctic Expedition of 1955–58. One of the original buildings erected during that expedition still stands, and houses memorabilia from the base's subsequent history. The base opened on 20 January 1957, and has been in constant use ever since. It has been considerably enlarged over the years, notably during an extensive rebuilding in 1976. Its eight interlinked pale-green buildings can accommodate up to 86 people in summer and about 10 each winter.

Hut Point is the site of the "Discovery" hut: a prefabricated wooden hut erected by Robert Scott's British National Antarctic Expedition of 1901–04. Meticulously kept records show that the building, which occupies 11.3 square meters (37 sq. ft), cost a modest 360 pounds 13 shillings and 5 pence sterling.

PACKED UP AND READY
New Zealand researchers set out from Scott Base with packed "Ski-Doos" for a three-day trip over the sea ice in spring. These motor toboggans are a cold but safe and efficient way to travel across sea ice.

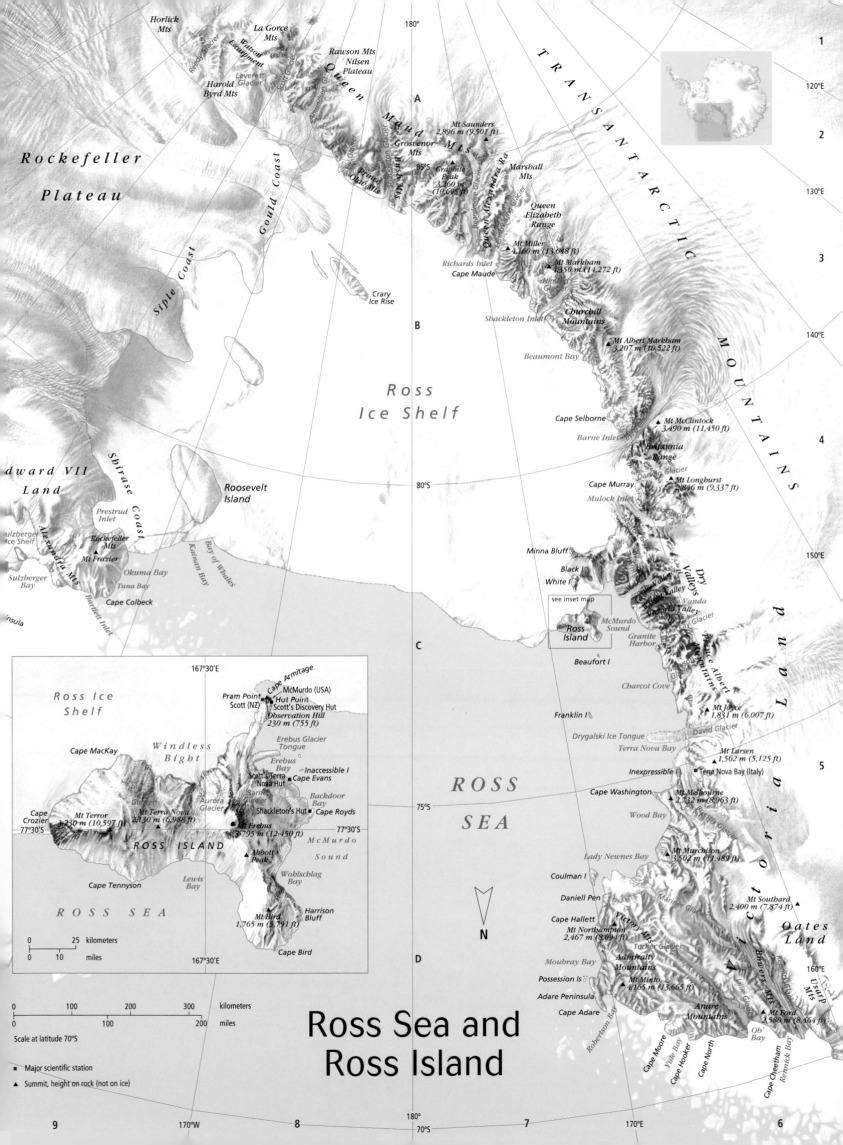

Ross Sea and Ross Island

Main map labels

1
Horlick Mts
La Gorce Mts
Watson Escarpment
Reedy Glacier
Leverett Glacier
Scott Glacier
Harold Byrd Mts
Rawson Mts
Nilsen Plateau

Rockefeller Plateau

Queen Maud Mts
85°S
Grosvenor Mts
Mt Saunders 2,896 m (9,501 ft)
Graphite Peak 3,260 m (10,695 ft)
Prince Olav Mts
Amundsen Glacier
Liv Glacier
Shackleton Glacier

Marshall Mts
Queen Alexandra Ra
Beardmore Glacier
Queen Elizabeth Range
Mt Miller 4,160 m (13,648 ft)
Mt Markham 4,350 m (14,272 ft)
Richards Inlet
Cape Maude
Nimrod Glacier

T R A N S A N T A R C T I C   M O U N T A I N S

120°E
130°E
140°E
150°E
160°E

Gould Coast
Siple Coast

Crary Ice Rise

Ross Ice Shelf

Churchill Mountains
Shackleton Inlet
Beaumont Bay
Mt Albert Markham 3,207 m (10,522 ft)

Cape Selborne
Mt McClintock 3,490 m (11,450 ft)
Barne Inlet
Britannia Range
Cape Murray
Darwin Glacier
Mt Longburst 2,846 m (9,337 ft)
Mulock Inlet
Mulock Glacier

dward VII Land
Shirase Coast
Prestrud Inlet
Sulzberger Ice Shelf
Rockefeller Mts
Mt Frazier
Alexandra Mts
Sulzberger Bay

Roosevelt Island

80°S

Minna Bluff
Black I
White I
see inset map
Ross Island
McMurdo Sound
Dry Valleys
Taylor Valley
Wright Valley
Victoria Valley
Vanda
Mackay Glacier
Prince Albert Mountains
Granite Harbor

Okuma Bay
Tuna Bay
Bay of Whales
Katnan Bay
Cape Colbeck
Bartlett Inlet

Beaufort I
Charcot Cove

C

Mt Joyce 1,831 m (6,007 ft)
David Glacier
Franklin I
Drygalski Ice Tongue
Terra Nova Bay
Mt Larsen 1,562 m (5,125 ft)
Terra Nova Bay (Italy)
Inexpressible I
Cape Washington
Mt Melbourne 2,732 m (8,963 ft)
Wood Bay

R O S S   S E A

75°S

Lady Newnes Bay
Mt Murchison 3,502 m (11,489 ft)
Coulman I
Daniell Pen
Mt Southard 2,400 m (7,874 ft)
Mariner Glacier
Cape Hallett
Mt Northampton 2,467 m (8,094 ft)
Tucker Glacier
Victory Mts
Admiralty Mountains
Moubray Bay
Possession Is
Adare Peninsula
Mt Minto 4,165 m (13,665 ft)
Cape Adare
Robertson Bay
Anare Mountains
Cape Moore
Yule Bay
Cape Hooker
Cape North
Cape Cheetham
Ob' Bay
Mt Ford 2,580 m (8,464 ft)
Bowers Mts
Lillie Glacier
Rennick Glacier
Usarp Mts

Oates Land

Victoria Land

70°S

Inset map

167°30'E

Ross Ice Shelf

Cape Armitage
McMurdo (USA)
Pram Point
Scott (NZ)
Hut Point
Scott's Discovery Hut
Observation Hill 230 m (755 ft)

Cape MacKay
Windless Bight
Erebus Glacier Tongue
Erebus Bay
Scott's Terra Nova Hut
Inaccessible I
Cape Evans
Backdoor Bay
Cape Royds
Shackleton's Hut

Cape Crozier
Mt Terror 3,230 m (10,597 ft)
Mt Terra Nova 2,130 m (6,988 ft)
Aurora Glacier
Barne Glacier
Mt Erebus 3,795 m (12,450 ft)
77°30'S

R O S S   I S L A N D

Abbott Peak
McMurdo Sound
Cape Tennyson
Lewis Bay
Wohlschlag Bay

R O S S   S E A

Harrison Bluff
Mt Bird 1,765 m (5,791 ft)
Cape Bird

Scale
0   25 kilometers
0   10 miles

167°30'E

Scale

0   100   200   300 kilometers
0   100   200 miles
Scale at latitude 70°S

■ Major scientific station
▲ Summit, height on rock (not on ice)

Grid / coordinates

180°
170°W
180°
170°E

9   8   7   6
1   2   3   4   5
A   B   C   D

N

RESEARCH STATION

Since 1956, the United States has had a full time research operation at McMurdo Station and from that first year it has been the largest base on the continent. Under the auspices of the National Science Foundation, a wide range of scientific research is conducted here, with significant logistical support.

FROZEN IN TIME

Still looking as if the men might walk back into the hut at any minute, Shackleton's hut at Cape Royds—like Scott's hut at Cape Evans—is frozen in both time and in reality. The dry air and cold temperatures have preserved the contents well.

BREAKING A HUDDLE

The Emperor penguin colony at Cape Crozier is the world's southernmost penguin colony and the first Emperor penguin colony ever discovered—in 1902 by members of Scott's first Antarctic expedition. Three members of his second Antarctic Expedition almost died in 1911 visiting the colony in the blackness and bitter cold of the Antarctic winter.

◀ PORT IN A STORM
From 1957 to 1964, Cape Hallett was the site of a year-round New Zealand/United States base that continued to operate as a summer station until 1973. After closure, many of the shelters were cleared away. Today, the main sign that humans have ever been here is this dome that was left standing to offer sanctuary in an emergency.

The Dry Valleys

The lunar landscape of Bull Pass between McKelvey and Wright valleys (below) is characteristic of the Dry Valleys terrain. The valleys receive about 1 centimeter (less than 1/2 in) of snowfall each year but evaporation greatly exceeds precipitation. The ground surface is dry, but there is permafrost beneath it. But the valleys are far from lifeless. In the 1970s scientists discovered minute bacteria, algae, and fungi living in tiny cracks in the rock, and blue-green algae have been found at the bottom of the few lakes that are not frozen solid. At the upper end of this simple food chain are three species of microscopic nematode worms that live on bacteria and can survive being freeze-dried in this extraordinary environment. The Dry Valleys is an environmentally fragile region, and human visits are carefully controlled; some parts of the Valleys constitute an environmental extreme, a place where plant and animal life can survive against incredible odds.

▲ THIRD CHOICE
Named after Scott's second-in-command, Cape Evans was the *Terra Nova* expedition's third choice, after Cape Crozier and Hut Point. The prefabricated hut was erected on the coastal slope of Mount Erebus. It was here that Shackleton's Ross Sea Party was marooned when the *Discovery* was swept away in May 1915.

◀ EXPLORING A CREVASSE
On Hut Point Peninsula, a New Zealander lowers himself into a large crevasse through a hole where the ice roof has collapsed. Crevasses are an ever-present danger in Antarctica. Snow will often collect on each lip, building outwards until the gap is bridged and the crevasse becomes invisible, and potentially very dangerous. Some snow bridges are thick enough to support a person or vehicle—and some are not!

East Antarctica

At the South Pole there is no east or west—only north. However, geographers divide the world into eastern and western hemispheres, and Antarctica matches these divisions. East Antarctica—also known as Greater Antarctica because most of Antarctica's landmass is here—lies below Africa, Asia, and Australia. It is characterized by an ancient and stable Precambrian rock shield covered by an ice sheet about 2 kilometers (1¼ miles) thick, and has a roughly semicircular shape that quite closely follows the Antarctic Circle. West (or Lesser) Antarctica lies south of the Americas between the International Dateline and the Greenwich Meridian,

and includes the Antarctic Peninsula, which is marked by the deep indentations of the Weddell and Ross seas and the extended arm of the Antarctic Peninsula. The Transantarctic Mountains form a natural barrier between East and West Antarctica.

There is a significant difference between the underlying landforms of East and West Antarctica although they are not immediately apparent because of the deep covering of ice across the continent. However, if the ice were to disappear, East Antarctica would be a continent similar in size and nature to Australia, whereas West Antarctica would be a much smaller land mass tapering away to a series of small islands and one large one with a long narrow spine along what is now the Antarctic Peninsula. In East Antarctica, large mountain ranges would stand revealed and there would also be a few large basins that would become virtually inland seas.

East Antarctica contains about 90 percent of Antarctica's ice. It has very little exposed land, virtually all of it along the coast, and it is generally occupied by wildlife and scientific bases in uneasy conjunction. Not many people have ever seen this remote region, and there are only a few weeks each year when even ice-breakers can approach. Until quite recently, exploration was largely along the shore, and the place names reflect the nationalities of early Antarctic expeditions. From west to east, the 11 major "lands" on this side of Antarctica are Dronning Maud Land, Enderby Land, Kemp Land, Mac.Robertson Land, Princess Elizabeth Land, Wilhelm II Land, Queen Mary Land, Wilkes Land, Terre Adélie, George V Land, Oates Land, and Victoria Land.

△ BARED BY THE SUN
Towards the end of summer, the maximum amount of underlying rock is exposed at the end of the snow melt. This is the 6.5-kilometer (4-mile) long Browning Peninsula at the southern end of the Windmill Islands, a biologically rich region close to Australia's Casey station.

▷ WELCOMING COMMITTEE
A meandering line of Emperor penguins strays from the main huddle to inspect an observer on the fast ice breeding platform of Auster Rookery, near Mawson station. By late summer, most chicks have safely fledged and are ready to head out to fend for themselves in the cold waters of the mighty Southern Ocean.

▽ ICEBOUND PERIMETER
Much of the coast of East Antarctica is unapproachable from the sea. The extent of the Windmill Islands, about 27 kilometers (17 miles) long and 10 kilometers (6 miles) wide is immediately apparent from the air. In this dangerous environment, aviation mishaps occur often, but surprisingly few have been fatal.

△ OBLITERATION OF A STATION

When the United States Wilkes station was built in just 17 days in 1957, it was located in a protective hollow. By the end of the first year it was being overwhelmed by snow accumulation and was largely abandoned in 1965. Unlike this wooden building that is still visible, many of the structures were Jamesway huts of fabric over metal that are now buried or collapsed.

The sub-Antarctic islands

The vast Southern Ocean and its neighboring oceans are far from empty. Islands are scattered across these seas—but there is a lot of water in between. Most of these islands were discovered by chance and few are habitable. Many early "discoveries" were mistakes, and over centuries new islands have been added to—and others deleted from—nautical charts. For someone standing on the icy deck of a ship and peering though driving sleet or rain, a distant iceberg seen vaguely through the fog may look a lot like land.

Even defining what constitutes a sub-Antarctic island is something of a challenge. Islands that are truly sub-Antarctic lie south of the Polar Front. Cool temperate islands of the southern oceans, including the Falkland Islands, Macquarie Island, and Iles Kerguelen and Crozet are covered here—all are important sites for Antarctic wildlife, and are frequently visited, significant locations in the human history of Antarctica. The comparatively warm temperate Auckland and Campbell islands, southeast of New Zealand, are also called sub-Antarctic in some contexts, —they are often visited on the way to the Ross Sea. Vegetation on sub-Antarctic islands ranges from mosses to small trees; life is gentler than on the Continent and the islands support a great variety of wildlife.

▲ REGAL BIRDS
Largest of penguins except the Emperor, King penguins live only on sub-Antarctic islands. The bird's Latin name, *Aptenodytes patagonicus*, is misleading because it is not found in Patagonia.

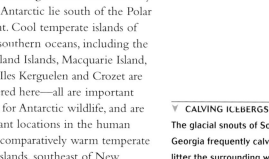

▼ CALVING ICEBERGS
The glacial snouts of South Georgia frequently calve and litter the surrounding waters with icebergs and ice fragments. Glacial retreat over the past 120 years has exposed expanses of bare rock that plants have yet to colonize.

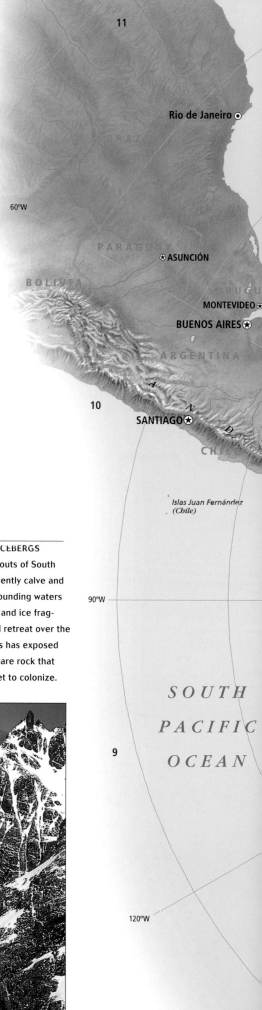

Rio de Janeiro

60°W

PARAGUAY
ASUNCIÓN
BOLIVIA
URUGUAY
MONTEVIDEO
BUENOS AIRES
ARGENTINA

10
SANTIAGO
CHILE

Islas Juan Fernández (Chile)

90°W

SOUTH

PACIFIC

9

OCEAN

120°W

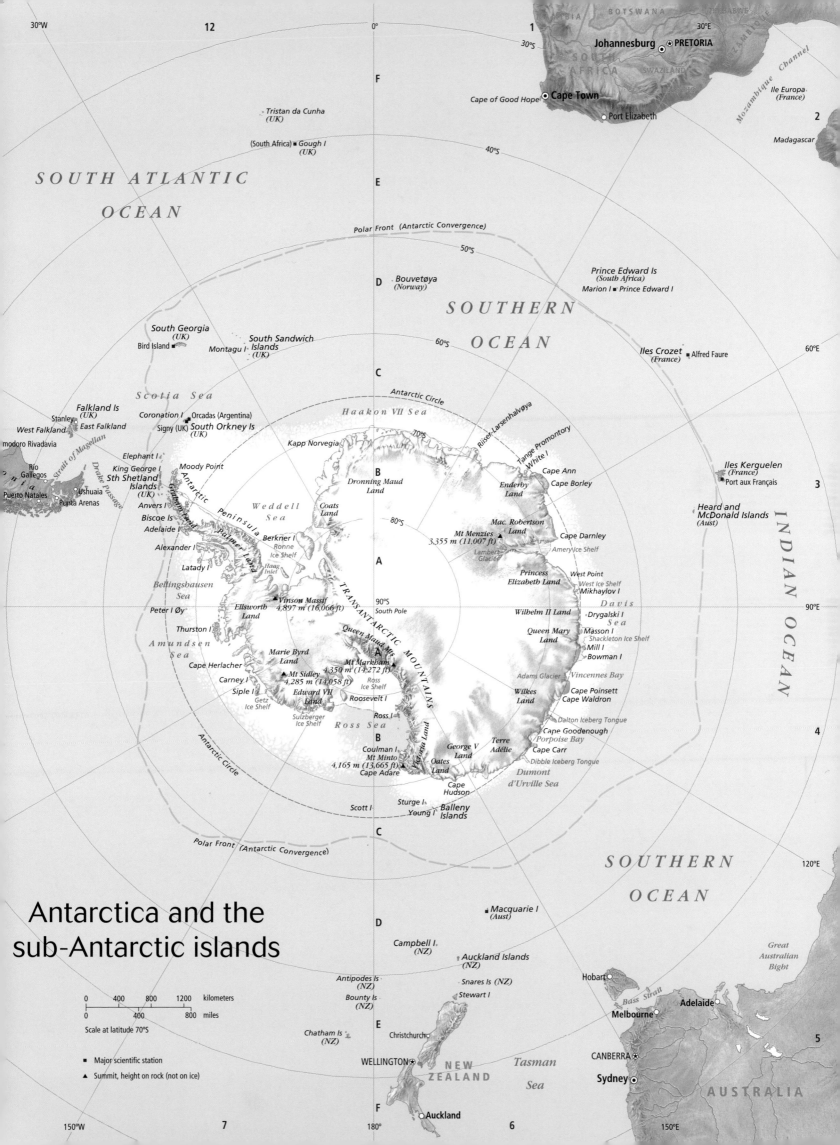

Antarctica and the sub-Antarctic islands

30°W   12   0°   1   30°E

F

BOTSWANA

ZIMBABWE

30°S
Johannesburg ⊙ PRETORIA
SOUTH AFRICA   SWAZILAND
MOZAMBIQUE

Cape of Good Hope ● Cape Town   Ile Europa (France)   2
● Port Elizabeth   Mozambique Channel

Madagascar

SOUTH ATLANTIC

OCEAN   E

Tristan da Cunha (UK)

40°S

(South Africa) ■ Gough I (UK)

Polar Front (Antarctic Convergence)   SOUTHERN

50°S

D   Bouvetøya (Norway)   OCEAN   Prince Edward Is (South Africa)   Marion I ■ Prince Edward I

South Georgia (UK)   South Sandwich Islands (UK)   60°S   OCEAN   Iles Crozet (France)   ● Alfred Faure   60°E

Bird Island ►   Montagu I ●

Scotia Sea   C   Antarctic Circle   Haakon VII Sea   Riiser-Larsenhalvøya

Falkland Is (UK)   Coronation I   Orcadas (Argentina)   Kapp Norvegia   70°S   Tange Promontory   Iles Kerguelen (France)

Stanley   Signy (UK) South Orkney Is (UK)   B   Dronning Maud   White I   ▪ Port aux Français   3

West Falkland   East Falkland   Elephant I   Moody Point   Land   Cape Ann   Enderby   Cape Borley

Comodoro Rivadavia   King George I   Antarctic   Weddell   Coats   Land   Mac. Robertson   Heard and McDonald Islands (Aust)

Río Gallegos   Sth Shetland Islands (UK)   Peninsula   Sea   Land   Mt Menzies   Land

Strait of Magellan   Anvers I   Graham Land   80°S   3,355 m (11,007 ft) ▲   Cape Darnley

Puerto Natales   Biscoe Is   Berkner I   Lambert   Amery Ice Shelf   INDIAN

Ushuaia   Adelaide I   Ronne   Glacier

Punta Arenas   Alexander I   Ice Shelf   A   Princess   West Point   West Ice Shelf

Drake Passage   Latady I   Haag Inlet   Elizabeth Land   Mikhaylov I   OCEAN

Bellingshausen   Vinson Massif   90°S   TRANSANTARCTIC   Wilhelm II Land   Davis Sea   Drygalski I

Sea   4,897 m (16,066 ft) ▲   South Pole   Queen Mary   90°E

Peter I Øy   Ellsworth   MOUNTAINS   Land   Masson I   Shackleton Ice Shelf

Land   Mill I   Bowman I

Thurston I   Amundsen   Queen Maud Mts   Adams Glacier   Vincennes Bay

Sea   Marie Byrd   Mt Markham   Wilkes

Cape Herlacher   Land   4,350 m (14,272 ft) ▲   Land   Cape Poinsett   4

Carney I   ▲ Mt Sidley   Ross   Cape Waldron

Siple I   4,285 m (14,058 ft)   Ice Shelf   Dalton Iceberg Tongue

Getz   Edward VII   Roosevelt I   Cape Goodenough

Ice Shelf   Land   Ross I   Porpoise Bay

Sulzberger   George V   Terre   Cape Carr

Ice Shelf   Ross Sea   B   Land   Adélie   Dibble Iceberg Tongue

Coulman I   Oates   Cape Hudson   Dumont

Mt Minto   Land   d'Urville Sea

4,165 m (13,665 ft) ▲

Cape Adare   Victoria Land

Antarctic Circle   Scott I   Sturge I   C   SOUTHERN

Young I   Balleny Islands   120°E

Polar Front (Antarctic Convergence)   OCEAN

Macquarie I (Aust)   Great Australian Bight

D

Campbell I (NZ)   Auckland Islands (NZ)   Hobart   Bass Strait   Adelaide

Antipodes Is (NZ)   Snares Is (NZ)   Melbourne

Bounty Is (NZ)   Stewart I   CANBERRA ⊙

E   Sydney ⊙

Chatham Is (NZ)   Christchurch   Tasman   AUSTRALIA

Sea

WELLINGTON ⊙   NEW ZEALAND

0   400   800   1200   kilometers
0   400   800   miles
Scale at latitude 70°S

F   150°E   5

■ Major scientific station

▲ Summit, height on rock (not on ice)

150°W   7   180°   6   ● Auckland

South Georgia

South Georgia has been likened to a piece of the Alps dropped in the middle of the ocean—a fair description, as the island is a summit of the Scotia Ridge, a line of mountains 4,350 kilometers (2,700 miles) long that links the ranges of the Antarctic Peninsula with the Andes through a submarine hairpin loop to the east. The South Sandwich and South Orkney islands are other exposed peaks along the ridge.

South Georgia has no near neighbors. Antarctica, South America, and Africa are about 1,500 kilometers (930 miles), 2,100 kilometers (1,300 miles), and 4,800 kilometers (2,980 miles) away, respectively, and the closest substantial landmass is the Falkland Islands, 1,400 kilometers (870 miles) away.

After the Falklands War, South Georgia ceased to be a Falklands' dependency, and in 1985 became the British Dependent Territory of South Georgia and the South Sandwich Islands. Despite the new title's equal billing,

South Georgia's land area—3,500 square kilometers (1,350 sq. miles)—is ten times larger than the 11 islands and several islets at the edge of the South Sandwich Trench that make up the South Sandwich Islands. These islands are a fog-enshrouded arc of very active volcanic, mostly ice-covered isles; they are rarely visited except by millions of breeding Chinstrap, Gentoo, Macaroni, and Adélie penguins.

South Georgia is uplifted rather than volcanic; its main landmass is surrounded by numerous offshore rocks and islets. Mount Paget, 2,934 meters (9,626 ft) high, is the tallest of a spine of mountains running northwest to southeast. Although Antarctic pack ice does not extend this far north, the bays around the coast freeze in winter and glaciers and ice fields cover some two-thirds of the island. The south is higher and more rugged than the north, and in many places the mountains fall sheer to the sea. Just four people live permanently on South Georgia.

LINKING CONTINENTS
There is little flat land on South Georgia—in most places the mountains tumble straight to the sea. Heinrich Klutschak (1848–90), an Austrian aboard *Flying Fish*, was the first to suggest that these mountains may be part of a mostly underwater ridge that forms a large loop connecting Antarctica and South America.

South Georgia lies 300 kilometers (186 miles) south of the Polar Front, where currents from the Weddell and Ross seas meet and swirl together to create a rich soup of nutrients. This attracts hordes of creatures from the top of the Antarctic food chain—penguins, albatrosses, fur seals, and whales.

King penguins breed prolifically on South Georgia: around 30 colonies accommodating over 200,000 pairs. Many more Macaroni penguins nest in the area (some 2.5 million pairs) but they are mainly found in large communities on inaccessible steep seaside slopes. About 120,000 pairs of Gentoo penguin raise their chicks on South Georgia.

South Georgia is also home to many flying birds. The largest of them, the Wandering albatross, has a resident breeding population of 4,300 pairs. A decline in their numbers has prompted a management plan at Albatross Island in the Bay of Isles to ensure that visitors do not disturb the birds. Other albatross species, including the Black-browed albatross, also nest on South Georgia. Light-mantled sooty albatrosses are found near Grytviken and Gold Harbor, while Gray-headed albatrosses frequent the northern end of the island. Petrels, skuas, gulls, terns, and shags can also be seen, as well as Speckled teal, and the South Georgia pintail. The latter, more obviously carnivorous than other species of duck, has been observed eating seal carcasses. The South Georgia pipit, a drab brown bird resembling a sparrow, is special. It lives only on South Georgia, but rats introduced from ships have drastically reduced its breeding range.

By the early twentieth century, seals had been hunted for their pelts to near extinction, but numbers have now risen again. More than 95 percent of the world's population of Antarctic fur seals is found on South Georgia and the little islands around it, and their numbers are continuing to increase by an average of 10 percent each year. In the weeks before Christmas, territorial adult males each jealously guard a stretch of beach. The males have departed by mid-January, but the remaining females rigorously enforce their private space for their pups. Smaller numbers of Leopard and Weddell seals are also found around South Georgia, and there are more than 300,000 elephant seals, although the huge dominant adult males leave the beaches early in summer.

It is thought that the population of fur seals has recovered very rapidly because the whales with which

the seals competed for food have also been slaughtered to near extinction. Whales breed much more slowly than seals, so whale populations have been slower to recover. Humpbacks, Minkes, and Right whales are sometimes seen in South Georgian coastal waters.

The Falkland Islands

Although north of the Polar Front and not, therefore, sub-Antarctic Islands, the Falklands are a popular stop en route to South Georgia. In 1982 this archipelago hit the headlines when British troops fought to retain ownership against invading Argentinian forces. Both nations invoke history to support their claims, but the Falkland Islanders are staunchly British.

Besides the large islands of West Falkland and East Falkland, there are at least 200 much smaller islands. The total population is about 2,500, some 1,500 people living in Stanley, the only town. The Falklands War has left a legacy of mined beaches and fields behind barbed wire and warning signs near Stanley—finds of live ammunition can be reported to the Ordinance Office.

West Point Island, on the western side of the group, is a convenient first landing site. Across the island, in a cleft high above the sea, Rockhopper penguins and Black-browed albatrosses nest in noisy profusion.

New Island, also on the west side of the group, is a nature reserve owned by expert naturalists Tony Chater and Ian Strange. The wildlife there includes fur seals, Gentoo, Magellanic, and Rockhopper penguins, Black-browed albatrosses, prions, and cormorants.

Carcass Island's wide range of bird life includes Magellanic and Gentoo penguins. Saunders Island's Port Egmont was the first British settlement in the Falklands (established 1766); some ruins from that era remain near the present owner's farmhouse. However, most visitors disembark in the north, where there are Gentoo penguins on one shore of the spit and Rockhopper and Magellanic penguins on the other. A colony of King penguins completes Saunders Island's array of flightless birds. On the grassy hill, where sheep graze among Magellanic penguin burrows, are Rockhopper rookeries and nesting pairs of Black-browed albatrosses.

A ROSY VIEW

Approaching Elephant Island, Ernest Shackleton wrote: "Rose-pink in the growing light, the lofty peak of Clarence Island told us of the coming glory of the sun." A humpback whale enjoys the view.

▼ FAR FROM HOME

This signpost at Bellingshausen station, on King George Island, plaintively reveals that home is a long way away: Murmansk is 16,500 kilometers (1,025 miles) in one direction, and Vostok station is 4,375 kilometers (2,718 miles) to the south.

South Shetland Islands

On the voyage from South America to the Antarctic Peninsula, the looming black volcanic crags and extensive glaciers of the more southerly South Shetlands Islands, such as Elephant Island so often shrouded in fog, provide a dramatic introduction to Antarctica. Many ships heading for the southern continent maneuver among the icebergs to make their first landfall on this part of the South Shetland archipelago.

Elephant Island lies at the northeastern end of the South Shetlands archipelago and is not on most itineraries. Even when conditions look good, ice in the bay and treacherous currents make landing tricky; it becomes well nigh impossible when seals and penguins crowd the beach. A more convenient location for wildlife-watching, Cape Lookout in the south, is frequented by many fur and elephant seals, and has a large population of penguins—Chinstraps, some Gentoos, and the occasional Macaroni.

Rounding the northern end of King George Island, the first accessible landing sites are Turret Point on King George Island and nearby Penguin Island. Currently about 8,000 breeding pairs of Chinstrap penguins and about half that number of Adélies hide away at Penguin Island's southern end. At the beginning of summer, the whole of this tiny island is usually still under snow. This snow cover and nesting Southern giant petrels can make it difficult to move on to the island from the landing site. Later in the season, when delicate mosses, grasses, and lichens are exposed, there may also be fur seals to contend with on the beach.

Across the narrow passage from Penguin Island, Turret Point is distinguished by a cluster of tall rock stacks at either end of a gravel beach. When the katabatic winds off nearby glaciers allow a landing, elephant and Weddell seals, Adélie and Chinstrap penguins, Kelp gulls, Antarctic terns, and Southern giant petrels may be observed. Nesting Antarctic shags crowd one of the offshore stacks.

Admiralty Bay is the largest harbor in the South Shetland islands, extending over 122 square kilometers

(47 sq. miles) and reaching depths of over 500 meters (1,640 ft). This anchorage was used by whalers in the early twentieth century; icebreakers are never needed to gain entry to it. Since 1996 the whole of Admiralty Bay has been an Antarctic Specially Managed Area. Many countries have bases here, including Chile's Teniente Rodolfo Marsh Martin station; one of the station's buildings sometimes operates as Antarctica's only hotel.

Going ashore on the tiny Aitcho Islands involves pushing through fringing kelp, but the resident wildlife is worth the effort. Large rookeries of Chinstrap penguins surround the most frequented landing site at the eastern end of the main island, while Gentoos nest among the bones at Whale-bone Beach. On the island's western side, elephant seals wallow near another landing beach but, after the snow has melted, footprints can all too easily damage the large beds of cushion moss, and the presence of humans may disturb nesting giant petrels.

Deception Island combines the novelty of sailing into an active volcano with the possibility of swimming, or at least wallowing, in thermally heated water. With the possible exception of Maxwell Bay on King George Island, Deception Island has little in common with the rest of Antarctica, and is dominated by derelict buildings and obsolete whaling works.

Port Foster comprises the whole flooded caldera within the island. Entering the caldera through a gap in the wall known as Neptunes Bellows, so named because of the violent wind that sometimes blows across the mouth of the entrance, is invariably exciting. Ships must negotiate a gap of less than 400 meters (1,310 ft) wide, hugging the northern side (where Cape petrels often gather) to avoid a submerged rock less than 2 meters (6½ ft) below the surface almost in the middle of the Bellows. The wreck of the *Southern Hunter*, a British whaler that ran aground avoiding a vessel of the Argentinian navy in 1957, lies on the southern side of the entrance.

Britain had claimed the island in 1908 as part of the Falklands Island Dependency, but its occupation of Deception Island came to a violent natural end in the late 1960s when volcanic eruptions wiped out both Chile's and Britain's stations. The British surrendered the wrecked base to the elements; the land around the abandoned Chilean station is now a Site of Special Scientific Interest, and Decepción station operates irregularly in summer.

Antarctic creatures find the water of Whalers Bay uncomfortably warm, and are reluctant to venture inside. In late summer a few fur seals, or even the odd Weddell or Leopard seal, may haul out on the beach, and there are always one or two Chinstrap or Gentoo penguins about. A full experience of Deception Island's wildlife, however, depends on the sea being calm enough for a landing at Rancho Point, known always by its unofficial name: Baily Head. Nevertheless, Baily Head is worth any effort to get there. Fur seals sometimes crowd the long, black sand beach, and multitudes of Chinstrap penguins make it their domain. The small valley near the high, rocky headland extends inland and becomes a two-way penguin superhighway in summer.

Other sub-Antarctic islands

The most easily accessible of the sub-Antarctic islands are the ones below Australia and New Zealand.

Long, thin Macquarie Island lies about 1,500 kilometers (930 miles) south of Tasmania, just north of the Polar Front. The ocean around the island became an Australian Marine Park in 1999. Like most sub-Antarctic islands, Macquarie Island has no trees; however, unlike other sub-Antarctic islands, the island is a plateau shaped by waves rather than ice. Australia's Macquarie Island station was built on a spit below Wireless Hill on 25 March 1948, and has since operated continuously. Affectionately known as Macca, the station houses about 20 people during winter and 40 in summer.

Campbell and the Auckland Islands are basaltic volcanic islands, covered in peat, that lie to the north of the Polar Front. They are cold, wet, and windy, but are ice-free (although shaped in the past by glaciers), and can support trees. Indeed, they have a wealth of vegetation that exists nowhere else, including many endemic vascular plants and the world's southernmost forests.

Both islands are National Nature Reserves; landing permits are required, and visitors must be accompanied by a representative of the Department of Conservation. Visitors to Auckland Island can inspect the old settlement site at Erebus Cove, Port Ross. But, most prefer nearby Enderby Island for the variety of rare wildlife there, and for the chance to walk through the mysterious rata forest with its bright red flowers.

Other sub-Antarctic islands are Peter I Øy (Peter I Island) Scott Island, Balleny Islands, Heard and McDonald Islands, Ilse Kerguelen, Iles Crozet, and Bouvetøya (Bouvet Island). DM

> **THE WESTWARD LIMIT**
The remains of a BAS Otter fuselage slowly deteriorate beside the aircraft hangar on Deception Island. The hangar marks the westward limit for visitors to Whalers Bay—the area beyond is protected to enable scientists to study plant re-establishment after eruptions and mudslides.

▼ **SCRAMBLING ASHORE**
The Iles Kerguelen have occasional sandy landing sites where penguins can emerge from the sea to breed. But the landscape is precipitous, and rough, rocky ground can pose a problem for penguins like these Kings, who cannot easily negotiate steep ground.

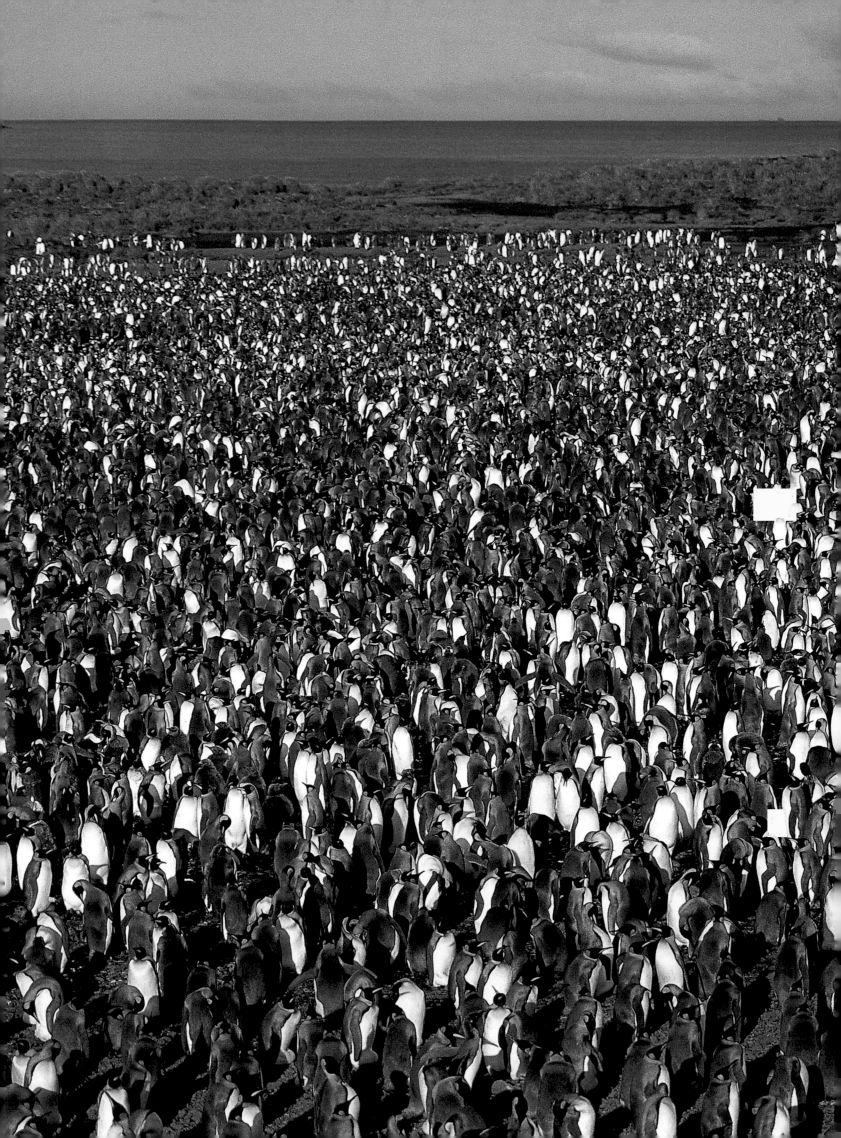

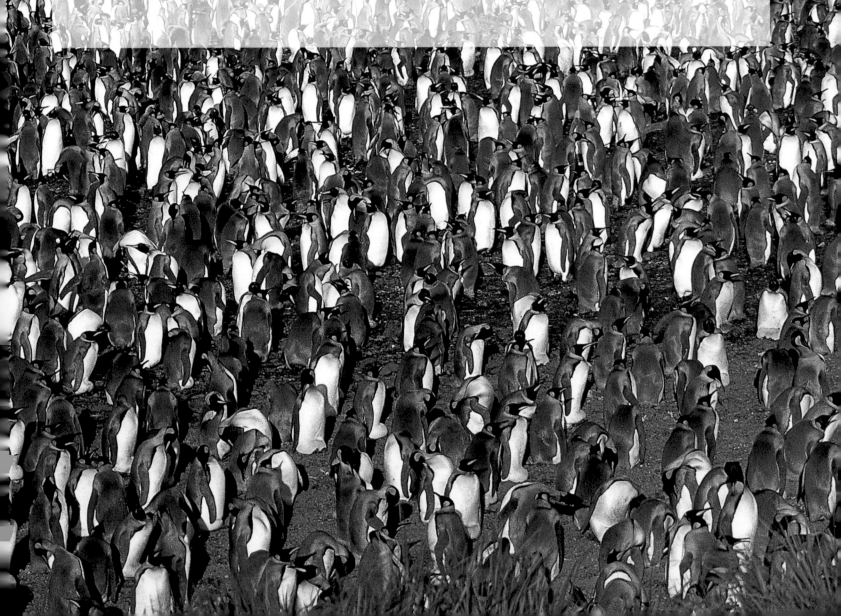

Part III
ANTARCTIC WILDLIFE

The southern continent is home to many remarkable species: the whales and seals of the Antarctic Ocean; the seabirds that must land to breed on the rocky shores and ice shelves of the frozen lands; the flightless penguins that must huddle for warmth on the Antarctic rock and ice. Plants find a precarious roothold in hostile soil, and tiny invertebrates and hardy mosses and lichens cling to life in a savage climate. All these life forms have evolved techniques to survive the challenging conditions that prevail in high latitudes.

Ecology
Plants and invertebrates

For land plants, life becomes harder the further south they grow. Plants can grow only where there is enough liquid water, and they must contend with freezing conditions, prolonged darkness, and extremes of light. As latitudes rise, plant life becomes less varied and abundant, with fewer species, fewer major plant groups, a lower biomass, a smaller proportion of ground covered by plants, and a shorter average plant stature. These changes resemble those along a moisture gradient from well watered regions into arid deserts—a closer analogy than it might seem, because the Antarctic region is extremely dry. As well, low temperatures and extreme fluctuations in light levels at high latitudes affect chemical processes such as photosynthesis, and plants are forced to take protective measures against these stresses.

The Antarctic regions

In terms of plant and animal life, the Antarctic region can be divided latitudinally into three distinct zones. The sub-Antarctic islands comprise six island groups near the Polar Front: Marion and Prince Edward islands; Iles Crozet; Iles Kerguelen; Heard and McDonald islands; Macquarie Island; and South Georgia. These islands support abundant but treeless vegetation, verdant in summer but brown when foliage dies in winter. The islands south and southeast from New Zealand (Snares, Auckland, Campbell, Antipodes, and Bounty islands) support similar vegetation, with the addition of upright shrubs and trees; they are in some contexts also called sub-Antarctic. All the major plant groups except cone-bearing trees (gymnosperms) occur in the sub-Antarctic. Microscopic invertebrates and larger invertebrates such as snails, earthworms, flies, beetles, and spiders are plentiful on the sub-Antarctic islands.

Maritime Antarctica consists of the west coast of the Antarctic Peninsula and its offshore islands, together with the islands south of the maximum extent of the sea ice: the South Sandwich, South Orkney, and South Shetland islands, and Bouvetøya (Bouvet Island). In this region, mosses and liverworts (bryophytes) are reasonably abundant and widespread, but only two species of non-woody flowering plants (soft-stemmed angiosperms) grow, and then only sparsely. Microscopic invertebrates and larger invertebrates such as springtails and free-living mites live in soil and amongst rocks and vegetation.

Continental Antarctica is the "Great Circle" of East and West Antarctica—the Antarctic Continent, with the exception of the west coast of the Peninsula. Near the coast in continental Antarctica, mosses, liverworts, lichens, and algae can grow, but are generally sparse. On the high Antarctic ice sheet at the Pole itself, there are probably no plants at all.

Microscopic invertebrates (protozoans, rotifers, tardigrades, nematodes, platyhelminths, and crustaceans) live where moisture is available on land and in lakes, but there are no land-based vertebrates in continental Antarctica: seals, penguins, and flying seabirds come ashore to breed, but they must live and feed at sea.

For plants, conditions in the ocean at high southern latitudes are much more favorable than on land. In the intertidal zones and shallow waters off sub-Antarctic islands, some of the world's largest marine algae form dense kelp communities of high biomass. Even in the colder shallow water off continental Antarctica, below the depth of the sea ice and where the sea bed is not scoured by icebergs, the biomass, stature, and diversity of marine algae are greater than that of land plants. Marine plants receive less light than land plants because light intensity is much diminished by passing through water, which suggests that low temperatures during the summer growing season and lack of liquid water are much more significant factors than low light intensity in the sparsity of Antarctic land vegetation.

Sub-Antarctic islands

The sub-Antarctic islands vary greatly in altitude and area covered by permanent ice. They range from unglaciated Macquarie Island, 433 meters (1,420 ft) high, and often fondly called "the great green sponge," to heavily glaciated Heard Island—2,745 meters (9,005 ft) at its tallest point. Their climate at sea level reflects that of the surrounding Southern Ocean. Average air temperature varies by only a few degrees from summer to winter, and the islands receive frequent precipitation as rain or snow/ice, so that plants have no extended periods without water.

From sea level to approximately 200 meters (650 ft) altitude, the ground is completely covered by herbaceous

▼ LIFE ON THE EDGE
In the dry, cold Antarctic environment, lichens (the circular dark patch on the nearest rock) benefit from the warmth absorbed and later released by dark rock surfaces. Snow melting at the rock surface provides much-needed moisture.

> INTERTIDAL GIANTS
Luxuriant beds of bull kelp (*Durvillaea antarctica*) are attached to rocky sub-Antarctic shores in the intertidal zone. Many invertebrates live amongst the holdfasts that attach these plants to the shore. On treeless sub-Antarctic islands, these marine algae are the largest plants.

flowering plants, bryophytes, and some ferns, except where there are rocky bluffs, screes, moraines, fresh lava flows, mobile sand sheets, seal and penguin disturbance, or landslides. The tallest vegetation, up to 2 meters (6½ ft) high, comprises tall tussock grass and large-leaved megaherbs (dicotyledons). In less favorable conditions there is short herb vegetation approximately 0.3 meters (1 ft) high: grasses and sedges (monocotyledons), small dicotyledons, mosses, or ground-hugging cushion plants.

At higher altitudes, lower temperature and increased exposure to wind reduce both the percentage of ground covered by plants and plant stature. On glaciated islands, ice sets a limit to plant distribution, but in recent decades the ice has been retreating and much new land has become available for colonization by plants.

With increasing altitude and abrasion, the proportion of flowering plants drops while the proportion of bryophytes rises. Cushion-form flowering plants like *Azorella selago* or *A. macquariensis* become most common—their tight mat of small, tough leaves resists wind and wind-driven abrasives, and their extremely long contractile roots, radiating shallowly in the soil, anchor them securely. Under the most extreme conditions, a surface of apparently bare gravel may be held together by a fine but dense, thread-like network of liverworts, their minute, reddish leaves sheltering in the crevices between the small stones.

Maritime Antarctica

The Peninsula forms a geographical and botanical transition zone between the far south of South America (Tierra del Fuego to Isla Hornos) and the great circular sweep of East and West Antarctica. Although the Drake Passage is a fearsome barrier, it is the narrowest ocean gap between Antarctica and other lands that plants might cross. The Peninsula is a mountainous, glaciated spine, essentially devoid of plants except near sea level, and most of its offshore islands are of similar physiography. The western fringe of the Peninsula near sea level is a relatively favorable environment for plants; the eastern fringe is colder.

Antarctica's native flowering plants and its only two vascular species—the small grass *Deschampsia antarctica*

> UNUSUAL ADAPTATIONS
Beneath the frozen surface of an Antarctic lake (right of illustration) is liquid water in which populations of micro-organisms live. Some photosynthesize, some ingest other organisms (heterotrophs), and fascinatingly some photosynthesize during summer when light is available but are heterotrophic during the dark winter.

and the small cushion plant *Colobanthus quitensis* (pearlwort)—are found on the western side of the Peninsula. By contrast, Tierra del Fuego, just to the north of the Drake Passage, has 417 species of vascular plants, of which 386 are flowering plants. Antarctica's two species of flowering plants are found as far as 68°S; at a similar latitude in the northern hemisphere, Iceland supports some 329 species, and a further 10 degrees north, 110 species of flowering plants grow in Spitsbergen. Suitability of soil and temperature and adequacy of moisture limit the distribution of Antarctica's grass and pearlwort. During recent climate changes, pearlwort populations have expanded—a trend that is likely to continue with future warmer summer temperatures.

On the eastern side of the Peninsula, and on Deception Island and Ross Island, volcanic activity enables some species of bryophytes to grow that are otherwise known only from further north. Most sites are unsuitable for temperate species, but it appears that there is a constant rain of plant fragments over Antarctica from

CUSHION PLANTS
Several species of *Colobanthus*, known as pearl-wort, grow as small, compact clumps on sub-Antarctic islands. *Colobanthus quitensis* is one of only two species of flowering plants growing in Maritime Antarctica.

warmer regions to the north, and some of these plants manage to become established on steam-warmed ground where the temperature is right and moisture is available from condensed steam.

Continental Antarctica
In East and West Antarctica plants are confined to snow-free and ice-free land, including mountains that project through the ice sheet, and to the land around the coastal fringes of the continent and of the Ross, Filchner, Ronne, and Amery ice shelves. In Victoria Land, vol-canically heated sites can support bryophytes. Apart from these areas, the only known plant forms in continental Antarctica are algae that grow on snow and ice, feeding on mineral nutrients from the snow and its trapped dust, and coloring the snow gray, green, orange, pink, or red.

Biologically, the most productive areas of Antarctica are the few small ice-free patches of land around the coastal fringe. These oases are breeding areas for marine animals, which carry ashore plant nutrients, particularly phos-phorus. Humans use the oases for building bases and for tourism, and so bring in pol-lutants and disrupt the local ecology.

FINDING A ROOTHOLD
The tiny tussocks of the grass *Deschampsia antarctica* grow amongst rocks and in soil-filled cracks in bedrock. Right to the southern limit of its distribution, *D. antarctica* can grow densely enough to form closed swards.

The invertebrates that live amongst Antarctic vegetation feed on dead plant material, soil bacteria, algae, and fungi. The carbon dioxide exhaled by the invertebrates and bacteria living amongst the stems of Antarctic moss microforests increases the rate of photosynthesis in the moss leaves, enhancing the production of sugars during the brief Antarctic summer growth period.

Lakes are common in ice-free areas, the best known being those in the Dry Valleys of southern Victoria Land and in the Vestfold Hills of East Antarctica. Each lake contains a population of micro-organisms: green algae, fungi, protozoa, and cyanobacteria and other prokaryotes (simple organisms that can survive in conditions too extreme for other forms of life, including volcanic soils with high concentrations of heavy metals, volcanically heated soils and water, and the extreme cold of Antarctic rock outcrops). The mix of organisms depends upon the water's chemical and physical properties: some lakes are far more saline than seawater, others are as pure as distilled water; some are well mixed, others are stratified, with stable layers of different temperatures, light levels, oxygen content, and other chemical properties. Some non-saline well-mixed lakes have underwater forests of moss more than a meter (3 ft) tall growing from the lake bed below the reach of the floating winter ice. The moss, *Bryum pseudotriquetrum*, is normally terrestrial, where it grows only a few centimeters tall. Sheets of photosynthetic cyanobacteria commonly form in thin layers on lake beds; if lithified (turned to rock), they would be called stromatolites. Dried fragments of these sheets, like pieces of torn paper, accumulate as lines of flotsam on lake shores.

The only plant communities known on the surface of the ice sheet are transient patches of snow algae, found at low altitudes where some summer melt occurs. But the whole ice sheet contains airborne microscopic spores of fungi and other single-celled organisms, such as bacteria, and no doubt tiny reproductive structures and fragments of plants. These organic bodies were probably trapped in snow. The distribution of organic particles deep in ancient Antarctic ice may prove to be a useful archive of data about earth's past environments. This great ice sheet also covers and seals lakes of liquid water, the largest being Lake Vostok. Fossil or living organisms occurring in its water and sediment await discovery.

The future of Antarctic life forms

High-latitude environments are undergoing change, but the consequences for polar organisms are difficult to predict. Global climate change is bringing warmer temperatures, changed precipitation, and changed ice cover. Depletion of the ozone layer over both the Antarctic and Arctic results in increased levels of ultraviolet light.

◄ COLOR FOR PROTECTION
The striking colors of Antarctic lichens—yellow, orange, and black—and of some Antarctic mosses—bronze and black—come from sun-screening pigments produced by the plants to protect their photosynthetic apparatus from damage by light of high intensity and by ultraviolet light.

On sub-Antarctic islands introduced animals are another threat; rabbits damage vegetation and promote soil erosion, cats prey on ground-nesting birds, and invertebrates accidentally introduced in building materials influence vegetation by consuming litter and live plants and compete with indigenous species. Toxic waste and fuel spills from Antarctic bases cause serious habitat destruction for Antarctic plants and animals. And, finally, tourism may become a problem: tourism is now ship-based, but serious environmental damage to plant and invertebrate habitats is inevitable if tourism ever becomes largely land-based. PS and DA

▲ TINTED SNOW
Unlike most Antarctic plants, snow algae are able to grow in snow itself, gleaning moisture from the snow as it melts. Various pigments in the cells of snow algae trap light energy for photosynthesis. En masse, the pigments color the snow—here, a pink hue.

Obtaining water in the driest continent

Although Antarctica's volume of ice is huge, low precipitation and considerable evaporation into the dry atmosphere make it the driest continent. Plants grow where liquid water is available during the summer, but the very low relative humidity dries them out quickly when there is no water.

Most vascular plants (flowering plants, conifers, ferns) try to keep their cells moist by having leaves covered with a waterproof cuticle, and by controlling variable openings in leaf surfaces, called stomates. For vascular plants to produce necessary sugars by photosynthesis, carbon dioxide gas, a raw material, must enter leaves via open stomates. But open stomates means water vapor can escape from moist cells inside to dry air outside, leaving the plant vulnerable to death by desiccation in the absence of a water supply.

Mosses and lichens have no cuticles or stomates and

desiccate quickly without water. But many mosses and lichens in Antarctica survive repeated cycles of wetting and drying by suspending metabolic activity when desiccated and resuming it when rehydrated. They take advantage of any available water and light—for example, when a dusting of snow over their bodies melts in midday warmth. Plants with this ability also exist outside the polar regions, and are called resurrection or poikilohydrous plants.

Life in the Southern Ocean

Identifying the food web

The first explorers to go to Antarctica, more than two centuries ago, were impressed by the abundant wildlife—whales, seals, penguins, and winged seabirds. The rich, ochrous colour on the underside of the sea ice also intrigued them. They recognized that it came from high concentrations of microscopic, single-celled plants called diatoms, which, they rightly concluded, represented the base of a rich food web.

From that time until less than 20 years ago, the Southern Ocean around Antarctica was thought to be one of the richest marine ecosystems on earth, with a simple food web in which diatoms fed shrimp-sized crustaceans called krill, and fish, whales, seals, and birds consumed the krill. We now know that this is a gross oversimplification. While the Southern Ocean has some extremely productive areas, it is no more productive, overall, than more nutrient-poor parts of the world's oceans, and its biology is just as complex as that of warmer waters. This ocean is really several interconnected ecosystems rather than a single large one.

Biological habitat

The extent of the Southern Ocean is not clearly defined because it takes in the southern parts of the Atlantic, Pacific, and Indian oceans. Depending on how its northern boundary is defined, the Southern Ocean accounts for 10 to 20 percent of the global ocean. It is dominated by the huge eastward flow of the Antarctic Circumpolar Current—the earth's largest current, which provides the main connection between the Atlantic, Pacific, and Indian oceans. Near the Antarctic Continent, the coastal current flows in the opposite direction, toward the west. The shape of Antarctica, with its large indentations, the Ross and Weddell seas, the northward-jutting Antarctic Peninsula, and the varying contours of the sea floor, change the breadth and speed of the circumpolar currents and give rise to localized currents. Oceanic boundaries between these currents and Antarctica's coastal features determine the sea-ice conditions and the level of biological activity. As a consequence, marine life around Antarctica has a very patchy distribution.

The Southern Ocean and its Antarctic Circumpolar Current developed after the break up of the super-continent Gondwana, some 40 million years ago. Fossil evidence indicates that the Circumpolar Current was established about 27 million years ago, and by roughly 22 million years ago the Polar Front (also known as the Antarctic Convergence) had formed. Sinking warmer northern water and colder southern water meet in the circumpolar zone. It is not uncommon for the surface water temperature to drop by about 2°C (35.5°F) across the Front. The Polar

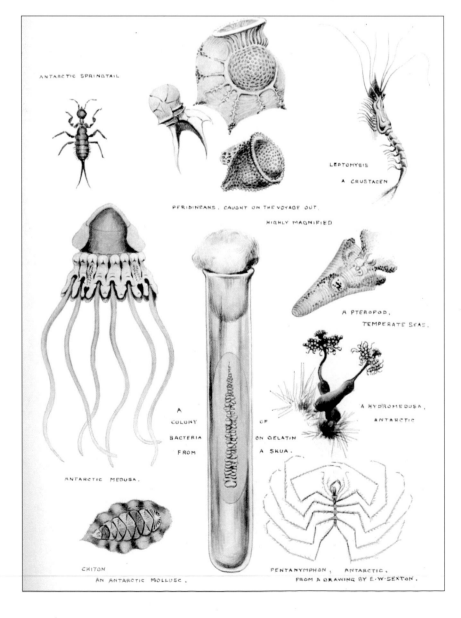

JOURNEY OF DISCOVERY
On Scott's inaugural journey to Antarctica in 1905, the explorers were delighted to discover a rich and unique fauna. These illustrations, from Scott's *The Voyage of the Discovery*, 1905, show some representatives of the invertebrate groups discovered on this expedition.

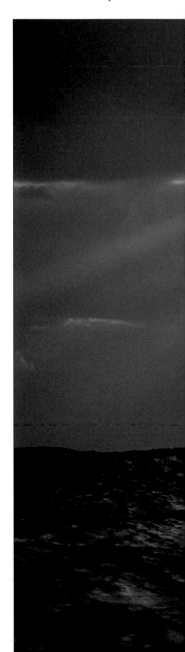

ALBATROSSES IN MOTION
These huge birds travel vast distances, apparently effortlessly gliding on long, narrow wings using the updrafts of the "Roaring Forties," a belt of strong westerly winds that occurs between about 40° and 50° south. Albatrosses are generally large and live for many decades. With their relatives, the petrels, they form one of two main groups of birds in Antarctica; the other group is the penguins.

Front is one of the earth's major oceanic boundaries and is a significant biological barrier.

The Southern Ocean, south of the Polar Front, can be divided into three zones that differ significantly in their biology, because of their environmental differences. The Permanently Open Ocean Zone (POOZ) is nutrient-rich but has relatively low levels of primary production. The main water flow is the Antarctic Circumpolar Current. The phytoplankton (single-celled plants) are generally tiny—nanoplankton, 2–20 micrometers in size. They are grazed by several groups of animals, but generally not by Antarctic krill (*Euphausia superba*).

The Seasonal Ice Zone (SIZ), south of the POOZ, is ice-covered in winter but is essentially open water in the summer months. It is the most productive zone of the Southern Ocean. Blooms of phytoplankton are common in the shallow, less saline water produced by the southward-retreating ice edge in spring and early summer. Large diatoms and blooms of *Phaeocystis* algae are the dominant phytoplankton, on which various groups of planktonic animals graze. Krill are abundant in this

◄ HUMPBACK DIVING
Humpbacks are readily identified by their low, broad blow, huge flippers, and also the way they arch their backs and show their flukes when diving. During summer they feed in Antarctic waters, and in early winter they migrate to their breeding grounds in coastal tropical waters. Populations were drastically depleted by whaling, and recovery is slow.

zone, especially in the more southern parts. Baleen whales, Crabeater seals, penguins, and other birds exploit the massive krill stocks.

The most southerly region, the Coastal and Continental Shelf Zone (CCSZ) is also known as the Permanent Pack Ice Zone—somewhat of a misnomer, as the region is not covered by pack ice at the time of its maximum retreat. The ice remaining is often fast ice. Phytoplankton blooms may be intense, but are generally short-lived.

Antarctic krill are uncommon, replaced by their smaller relative, *Euphausia crystallorophias*. In the absence of krill, birds and mammals are less abundant than in the SIZ.

The temperature at which seawater freezes depends principally on its salinity. Around Antarctica, the freezing point of seawater is close to −2°C (28.4°F), and it is rarely warmer than 4°C (39.2°F). As a result, Antarctic marine organisms experience constantly low temperatures.

For most of the year, the surface of the sea around Antarctica is frozen to a depth of up to several meters, but with an average thickness of less than a meter (3 ft). The freeze begins in March and reaches its maximum extent in September, when the sea ice covers almost 20 million square kilometers (8 million sq. miles)—which is nearly two and a half times the area of Australia. The sea ice is a major habitat for microscopic organisms, including the

▼ SHARING THE SEAS
Both albatrosses and Dusky dolphins have a circumpolar distribution that extends north of the Antarctic sea-ice zone. Albatrosses feed mostly on prey found at the surface of the sea, whereas Dusky dolphins eat fish and squid.

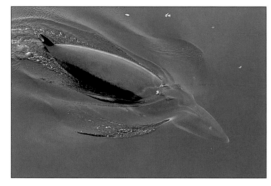

◄ SMALL BUT NUMEROUS
Minkes are the smallest and most numerous of the baleen whales, and their numbers seem to be increasing. Unlike most baleen whales, their blow is low; it has a distinctly fishy smell.

▼ BIOLOGICAL ZONES
The Polar Front is a major oceanic boundary in the Southern Ocean between the warmer northern waters and the cold Antarctic waters. South of the Polar Front is an ice-free zone surrounding the Seasonal Ice Zone, which is ice-covered for most of the year. Close to the coast there are regions that are permanently ice-covered.

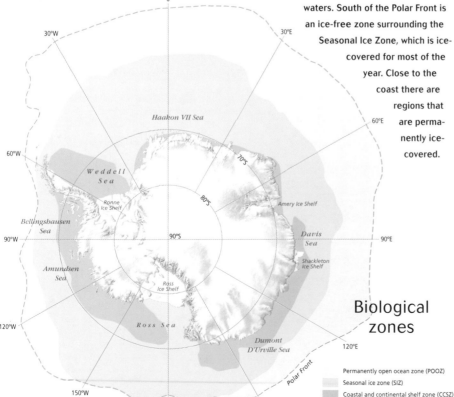

Biological zones

Permanently open ocean zone (POOZ)
Seasonal ice zone (SIZ)
Coastal and continental shelf zone (CCSZ)

diatoms seen by the early explorers. They proliferate on its underside, as well as at the boundary between the sea ice itself and the snow that settles on its surface, which depresses this boundary into the water. Cracks and other features in the ice are refuges for the small animals that graze the microorganisms. In addition, the sea ice is a platform on which some species of seals and penguins breed.

Algae and its associated microbial community of protozoa (single-celled animals) and bacteria on the underside of the sea ice is effectively the only food available to grazers while the sea is covered with ice, which can be for nine or ten months of the year. The under-ice algae on fast ice (that is, sea ice attached "fast" to the Antarctic continent) supports crustacean grazers—copepods, krill,

and amphipods—as well as fish. The algae living on the underside of pack ice (which is sea ice found further offshore than fast ice that drifts with the currents) is grazed by krill. The distribution of sea-ice algae and their associated microorganisms is extremely patchy.

As the sea freezes in autumn and winter, cold, highly saline water is excluded from the forming ice, which has a lower salinity than that of the seawater from which it is produced. The cold, saline water sinks to the sea floor and flows north into the northern hemisphere, as Antarctic Bottom Water (ABW). This is one of the main processes that create the vertical circulation and mixing of the global ocean.

During spring and summer, melting sea ice in the SIZ produces a layer of less saline surface water. Phyto-plankton bloom in this nutrient-rich, stable, shallow layer, which receives 24 hours of sunshine and produces localized areas of high output, capable of sustaining the biological diversity and richness that so impressed the first explorers of this region.

There is serious concern that global warming may reduce the amount of sea ice around Antarctica, thus diminishing the production of ABW. Mathematical models indicate that a lessening in the vertical mixing of the oceans could lead to stagnation of Southern Ocean bottom water, depleting it of oxygen and so making it unin-habitable by most living things. A decline in the amount of sea ice would also mean less habitat for the organisms that use it, and lead to a reduction of the relatively less saline water that promotes the ice-edge blooms.

▲ IN SEARCH OF FOOD
Adélie penguins breed on and around the Antarctic Continent. They will travel 200–300 kilometers (120–190 miles) away from their nesting sites in search of food—krill and other small crustaceans, and also fish. Their breeding success rate is poor during years when the sea ice is extensive.

▲ ALGAL BLOOMS
Shield-like scales cover this
single-celled alga (about
15 micrometers across). It
occurs throughout the world's
oceans where the water temperature is above 2°C (35.5°F).
It can be so abundant that the
sea appears milky white, or
so extensive that blooms are
visible from space.

Top

▲ CELL SCULPTURES
The lorica, or house, of this
single-celled tintinnid is
40 micrometers long. It is
secreted by the animal and
decorated with the scales and
wall fragments of its prey.
Tintinnids attract prey items
by beating minute, hair-like
structures, called cilia, to
create a water current.

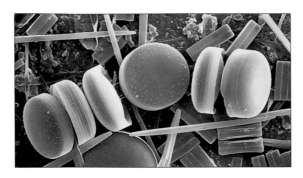

◄ DIATOM DISKS
Diatoms are the most abundant phytoplankton in the sea around Antarctica.
This scanning electron micrograph
shows cells of the diatom *Porosira pseu-
dodenticulata* (50 micrometers across).
When these cells divide, they form short
chains that are easily broken up. The
glassy walls of silica accumulate as thick
deposits on the bottom of the sea.

Protists

At the level of the single cell, the distinction between
plant and animal becomes clouded. These days they are
grouped together and called protists.

Phytoplankton are microscopic, floating, single-celled
plants. The 200 or so species identified from Antarctic
waters are startlingly diverse in size, shape, lifestyle, and
food value to grazers. The smallest measure about
1 micrometer (1 thousandth of a millimeter), and they
vary in shape from nearly spherical to hair-like, up to
about 4 millimeters (about $\frac{1}{8}$ in).

Cells of one group of phytoplankton, the diatoms,
are encased in intricately sculptured, glassy walls; others
are covered with finely patterned scales. Some can
swim; others drift enclosed in mucus.

As well as these single-celled plants, there are single
celled animals (protozoa) that feed on the phytoplankton, bacteria, and detritus in the water. They, too, are
highly diverse in shape, size, and lifestyle. Some filter
their food from the water, others glide over surfaces
grazing as they go, and others are voracious hunters.
Some plants are able to hunt for food—an ideal adaptation in Antarctica, where light levels are too low for
photosynthesis for much of the year. At the other
extreme are protozoa, which retain the photosynthetic
systems from the phytoplankton they consume, and use
their photosynthetic products as their food. Grazing by
plants and photosynthesis by animals is called mixotro-
phy. Protist grazing removes a substantial amount of the
bacteria and smaller phytoplankton. Bacteria feed on the
dissolved organic material that is produced by all organisms, particularly the phytoplankton. As protozoa are
eaten by krill and other grazers, they and the bacteria
represent another pathway in the food web between
phytoplankton and the grazers. This pathway is called the
microbial loop.

Bacteria and viruses

As elsewhere in the world's oceans, bacteria are a major
component of the Southern Ocean food web. Their
concentration is about a billion per liter (2 pints), and
they break down detritus, remobilizing the nutrients. In
addition, they play several other roles that can have
ecosystem-wide ramifications. Thus bacteria are crucial
to marine biological processes, playing a pivotal role in
the grazing food chain, the microbial loop, particle sinking, and carbon fixation and storage.

The smallest biological entities in the sea are viruses.
Viruses consist of genetic material (DNA or RNA)
enveloped in a protein coat. They are not living, in the
sense that they cannot metabolize, or reproduce by
themselves. They infect a host cell, hijack its reproductive system, and produce masses of new viruses which,
when the host cell bursts, can infect yet more host cells.
Their abundance in the Southern Ocean is similar to
that elsewhere in the sea, about 10 billion per liter
(2 pints). The roles these viruses play are only now being
explored. They probably infect all types of marine
organisms, and influence the species composition of
microbial communities and nutrient cycling, and are
involved in the transfer of genetic material between host
organisms. Virus infection accounts for about half of the
mortality of marine bacteria, and may inhibit phytoplankton production by up to 80 percent. Viral destruction of bacteria may be one of the major sources of dissolved organic matter in the sea, which is in turn eaten
by bacteria.

The importance of microscopic organisms in the sea
has become increasingly clear over the past two decades.
It is now known that much, often more than half, of the
total flux of matter and energy passes through the microbial loop. The contemporary view is of an organic matter
continuum from small simple molecules of amino acids
(the building blocks of proteins) and sugars through
colloids to transparent polymer particles, called oceanic
"dark matter". Recently, it has been proposed that these
organic constituents of seawater interact to form a very
dilute gel with a structure that has microscale hotspots,
which are sites of heightened microbial activity. Micro-
organisms, like larger plants and animals, do not live by
themselves but in communities with other organisms,
each having its own role.

Krill and other grazers

Krill—food for fish, birds, and mammals, and the target
species of a commercial fishery—are one of the central
elements in the Southern Ocean ecosystem. There is
probably a greater mass of Antarctic krill (*Euphausia
superba*) than any other single species on earth. Krill has
a patchy, offshore, circumpolar distribution. Abundance
and distribution vary from year to year, in some cases
with profound consequence for its predators.

Krill are a group of about 80 different species of crustaceans, the euphausiids, which resemble shrimps, and live
in the open ocean. In the Antarctic context, the term krill
refers to Antarctic krill, the largest and the most abundant
of the five species of euphausiids living in the Southern
Ocean. The dominant euphausiid in Antarctic coastal
waters is *Euphausia crystallorophias*, so-called Crystal krill,
a much smaller species than *E. superba*.

Very little is known about krill. Estimating their abundance is difficult because of their patchy distribution and their ability to avoid capture in research-size nets. The latest estimate is between 60 million and 155 million tonnes. Some studies indicate there may have been declines in some areas over the past 20 years or so, but whether these declines are part of a cycle is not known.

Krill characteristically form swarms in which the concentrations of animals can be about 30,000 individuals per cubic meter (1⅓ cu. yards); an easy target for whales, seals, and penguins—and fishing trawlers. Krill mainly eat phytoplankton, which is scarce for much of the year because of the wintertime absence of sunlight for photosynthesis. Laboratory studies have shown that Antarctic krill can cope with long periods of starvation.

Because of the ecological importance, resource potential, and biomass of krill in the Southern Ocean, other zooplankton tend to receive much less attention. These include other euphausiids, copepods, salps, amphipods, chaetognaths (arrow worms), and larval fish. Of these, copepods and salps can reach high abundances and play key roles. Non-krill zooplankton have been reported to account for 75 percent of the total zooplankton biomass in the Scotia Sea, north of the Antarctic Peninsula, and copepods have been found to constitute 60–87 percent of the total summer zooplankton biomass in Admiralty Bay on King George Island. Recent studies show that the contribution of krill to the zooplankton biomass in the Southern Ocean has been overestimated. It has become apparent that there is dramatic variability in the distribution and abundance of zooplankton geographically, as well as seasonally and interannually.

Copepods, like krill, are also crustaceans. More than 70 species are recognized in Antarctic waters. Generally microscopic, only a few species are larger than 1 millimeter (1/32 in). Some species graze on protists, others are carnivores, yet others are omnivorous. Estimates based on the abundance of copepods and their energy requirements indicate that they may consume more than eight times as many protists as krill. Unlike krill, with only few exceptions, copepods are not the principal food of the top predators. Instead, they are grazed by fish, including the so-called Antarctic silverfish or herring (*Pleurogramma* species), which, in turn, are eaten by seals, penguins, and other seabirds.

Salps are planktonic tunicates (sea squirts). They feed by pumping water through a fine mucus net filter, and are able take in a wide size range of particles. They are delicate, gelatinous organisms with a life history alternating between solitary organisms and colonies that reproduce asexually. The production of colonies coincides with the springtime phytoplankton bloom. As a solitary salp can produce hundreds of colonies, high concentrations of these organisms can develop rapidly in spring and summer in response to food availability. Their high, localized abundance and high filtration rates enable

▲ AS COMMON AS KRILL
The Antarctic krill (*Euphausia superba*) could be the most abundant single species on earth. Krill graze on microorganisms and are themselves food for whales, seals, birds, and fish. They gather food with their front legs, which form a feeding basket, and swim with their rear legs. This species is 50 millimeters (2 in) long.

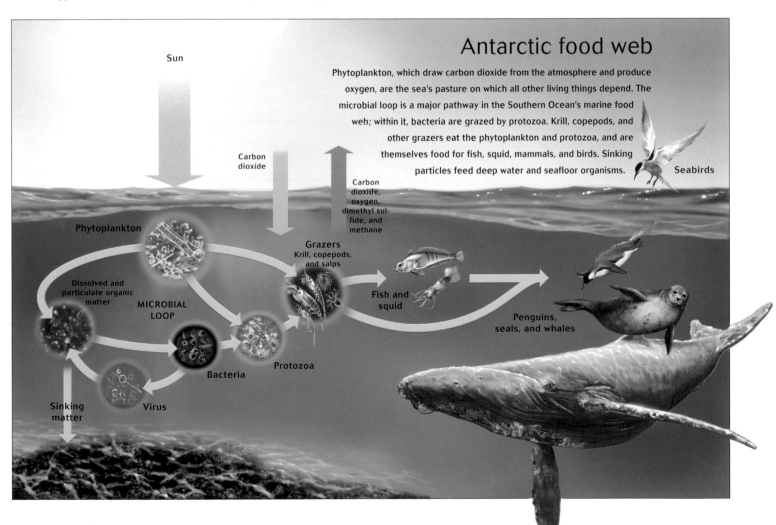

Antarctic food web

Phytoplankton, which draw carbon dioxide from the atmosphere and produce oxygen, are the sea's pasture on which all other living things depend. The microbial loop is a major pathway in the Southern Ocean's marine food web; within it, bacteria are grazed by protozoa. Krill, copepods, and other grazers eat the phytoplankton and protozoa, and are themselves food for fish, squid, mammals, and birds. Sinking particles feed deep water and seafloor organisms.

Sun

Carbon dioxide

Carbon dioxide, oxygen, dimethyl sulfide, and methane

Seabirds

Phytoplankton

Dissolved and particulate organic matter

MICROBIAL LOOP

Bacteria

Virus

Sinking matter

Protozoa

Grazers
Krill, copepods, and salps

Fish and squid

Penguins, seals, and whales

them to consume 10–100 percent of the protists in these areas. Thus salps are able to reduce concentrations of food for other planktonic grazers. These observations, plus finding that areas dominated by krill are usually essentially devoid of salps, and that areas in which salps are abundant are low in krill, have led to a reappraisal of the ecological role of salps in the Southern Ocean.

Investigations indicate that krill reproduction and the survival of larvae are also affected by blooms of salps. When sea ice is less extensive, salps, rather than krill, are more abundant. At least in the region of the Antarctic Peninsula, there is evidence, spanning the past 50 years, of fewer and fewer winters having extensive sea ice. This decrease in sea ice correlates with the increase in air temperatures over this time. As decreased krill abundance is related to a decrease in sea-ice extent, there is concern about the food availability for krill predators.

Although salps, unlike krill, are not a major constituent in the diet of vertebrate predators, they play another important role in the Southern Ocean. As their feces sink quickly, they contribute to the transport of carbon from surface waters to the deep ocean. This vertical flux of carbon in the world's oceans is one of the major pathways in the global carbon cycle—atmospheric carbon dioxide, taken up by phytoplankton, is transported to the deep ocean, where it is only slowly recycled back to carbon dioxide by the activity of deep-sea organisms.

Antarctic fish

More than 20,000 species of fish occur worldwide, but only about 120 of them are found south of the Polar Front. This Front represents a major oceanic barrier to the movement of fish species, especially those that live in coastal and shallow habitats. Also, the Antarctic continental shelf is deeper than the shelf around other continents, and has no shallow connection to these continental shelves. This isolation of Antarctic fish has played a major role in their evolution and in the composition of Antarctic fish communities. For example, many of the fish that live on the Antarctic continental shelf have adaptations found in deepwater fish from other parts of the world. Unlike other parts of the world's oceans, Antarctic fish are uncommon in surface waters.

Fish that live in Antarctic coastal bottom waters are by far the most diverse and abundant. In marked contrast to other parts of the global ocean, the Southern Ocean contains extremely few true schooling pelagic (open-sea) fish. The sea-ice zone is the habitat of adults of many of these, but they spend their first years of life in deep water. The Antarctic herring or silverfish (*Pleurogramma* species) is pelagic and often, especially as juveniles, associated with sea ice. It feeds on copepods,

euphausiids, and juvenile fish, and is, in turn, eaten by marine mammals and birds. Lacking a swim bladder, it achieves neutral buoyancy with fat deposits. *Pleurogramma* reaches a length of about 25 centimeters (10 in) and is an abundant and important organism in the Antarctic marine food web. Myctophids (lantern-fish), 2–30 centimeters (¾–12 in) long, are the other important pelagic group of fish in Antarctic waters. They are called lanternfish because of the groups of luminous organs (photophores) on their head and body. They are opportunistic feeders, consuming copepods, euphausiids and other grazing species, as well as fish eggs and juvenile fish. During the day, lanternfish can be found at depths of about 200 meters (650 ft) but migrate to be near the surface at night. They are unusual among Antarctic fish in that they have a swim bladder.

Unlike most Antarctic invertebrates, body fluids of Antarctic fish are less saline than seawater—their freezing point depression is about −1°C (30.2°F). They have all evolved antifreeze compounds called glycoproteins, which depress the freezing point of the water in their body fluids. These antifreezes work by binding to the surface of forming ice crystals, thus preventing their growth.

Most Antarctic fish are less than 30 centimeters (12 in) long, but a few species grow to over 1.5 meters (5 ft) long and weigh more than 50 kilograms (110 lb). Antarctic fish generally grow slowly, take three to eight years to become sexually mature, and have long life spans and low metabolic rates. They produce only a few large, yolky eggs, so their reproduction rate is generally low. These characteristics put Antarctic fish at serious risk of overfishing. Marbled rock cod (*Notothenia rossii*), Mackerel icefish (*Champsocephalus gunnari*), and Patagonian toothfish (*Dissostichus eleginoides*) have all suffered the predictable pattern of exploitation—initially high catches that rapidly decline to an uneconomic point, whereupon fishing effort is transferred to a new target species, and the pattern is repeated.

Antarctic squid

Squid are almost certainly one of the most important components of the Southern Ocean ecosystems, yet they remain one of the big unknowns. Because of their ability to avoid capture in nets, very little is known of their abundance, longevity, and lifestyles. Most species that are netted are not the squid taken by toothed whales. Also squid caught in nets are much smaller than those found in dietary studies of whales. Squid form a major part of the diet of birds, particularly albatrosses, seals, and toothed whales (except Orcas). Squid eat voraciously, some species consuming up to 30 percent of their body weight a day; myctophids and deepwater fish are their major prey. As these fish feed mainly on copepods, it is emerging that squid may represent an important food web link between copepods and toothed whales, seals, and birds.

Benthic communities

Just as the abundant wildlife impressed the explorers of Antarctica, so did the organisms that live on the sea floor, brought up in grabs, dredges, and nets. The coastal and

▲ LIFE'S CYCLE

An Adélie penguin feeds its chick. Foraging by penguins and other seabirds can have a major impact on prey species near their breeding colonies. They also have a substantial influence on the terrestrial environment of their breeding areas. Nest-building activities modify the landscape, and the excrement, egg shells, feathers, and carcasses of dead birds are major sources of nutrients for plant growth.

continental shelf areas of Antarctica have rich benthic (bottom-dwelling) communities, generally dominated by animals, which feed on particle of matter raining down from the surface waters.

One of the best-studied benthic environments at high latitude in Antarctica is McMurdo Sound. Generally the area, regularly scoured by ice down to a depth of about 15 meters (50 ft), is almost devoid of algae. Any animals there are mobile foragers, such as sea urchins, starfish, various worms, crustaceans, and fish. However, below this zone is a rich and diverse flora and fauna. In shallow coastal environments, benthic microalgae make a substantial seasonal contribution to primary production. Some 700 species of seaweeds have been reported from Antarctic waters, of which about 35 percent are endemic. The zone below that scoured by ice is colonized by a diverse array of filter feeders, including anemones, soft corals, molluscs (shellfish), ascidians (sea squirts), and tube worms. Here too are mobile scavengers, including starfish, sea urchins, pycnogonids (sea spiders), and fish. Below about 30 meters (100 ft) depth, the dominant animals are sponges covering up to 55 percent of the bottom in McMurdo Sound. These sponges vary widely in size and shape and grow only very slowly. The largest are shaped like a volcano, and are up to 2 meters ($6\frac{1}{2}$ ft) tall and 1.5 meters (5 ft) across. This sponge community provides homes for many mobile species, and anemones, tube worms, bryozoans, molluscs, and ascidians also occupy this zone.

Some sponges' skeletons are composed of silica fibers called spicules that provide support for the sponge and deter predators. Some Antarctic sponges contain algae that provide them with nutrients. More recently, it has been discovered that the glassy spicules on the sponges also function as optical fibres, transmitting light to the algae that adhere to the spicules deep within the sponge.

Areas around Antarctica where the sea floor is sandy or muddy contain burrowing animals. The densities of these animals are among the highest found anywhere on earth.

As is the case in temperate and tropical waters, predator–prey interactions are thought principally to determine the composition of Antarctic benthic communities. However, in Antarctic waters, icebergs scour the continental shelf, which has a depth of about 500 meters (1,640 ft). Statistically, this scouring would be expected to occur at least once every 230 years, though depth and proximity to active iceberg calving areas would mean much more frequent scouring of some parts of the continental shelf. Areas scoured by icebergs look like ploughed fields with parallel grooves, some bordered by raised embankments. The passage of the iceberg strips all organisms from the bottom. There is a sequence of recolonization of iceberg scours with mobile animals, including fish, sea urchins, and mollusks. They are followed by animals attached to the bottom, including tube worms and sea squirts.

Habitat destruction by iceberg scouring and subsequent recolonization may be a principal cause of the great differences in community composition between habitats on the Antarctic continental shelf.

In addition to the seaweeds and large bottom-dwelling animals, the Antarctic sea floor is covered with microscopic algae, protozoa, and other animals and bacteria. The protozoa and bacteria also live in benthic sediments. The benthic microalgae are generally subdivided into two groups: those that live in the top few layers of sediment, and those called ephiphytes that live attached to other living things, such as seaweeds, or on the surface of rocky seafloors. These algae can be extremely abundant, especially in some habitats, such as beds of sponge spicules, where the surface area of the sediment is increased dramatically by the long, needle-like spicules to which the algae are attached. In addition, the spicules discourage grazing animals. The benthic microalgae are thought to make a significant contribution to primary production in shallow coastal areas and play an important role in providing nutrients to the benthic animals, as well as contributing to the organisms living in the overlying water column.

Benthic bacterial concentrations are usually many times higher than the concentrations in the overlying water column, but similar to those found in sediments in other parts of the world. Bacterial activity in sediments is usually confined to the top 5 centimeters (2 in). The rates of bacterial processes in Antarctic marine sediments in breaking down organic material and recycling it back to carbon dioxide are thought to be similar to that in the deep oceans of the world. To better understand the role of the Southern Ocean in the global carbon cycle, much more research is required to measure the rates of benthic microbial activities. HM

▲ DINNERTIME FOR SEALS
A Weddell seal devours its prey, an Antarctic cod. These seals live further south than other seals. They tend to stay close to their breeding sites, feeding on a wide range of prey items.

▼ FIERCE PREDATORS
The seabed communities of Antarctica are surprisingly diverse and colorful. Many species of starfish live in coastal areas, and some, such as this *Odontaster validus*, are found in large feeding aggregations. Voracious predators, they will prey on other starfish much larger than themselves.

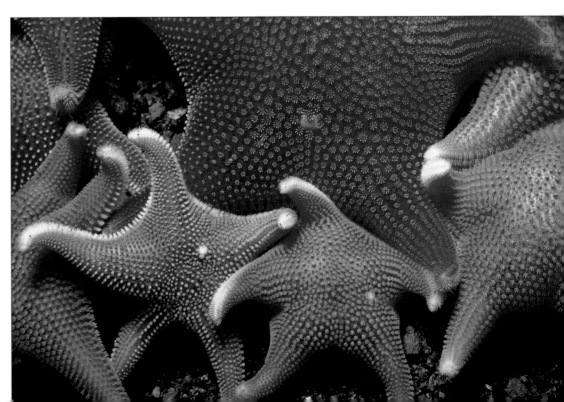

Mammals
Seals

Seals are a significant and highly visible part of the wildlife of Antarctica. They are classified into three families: Phocidae, the true seals, or hair seals; Otariidae, the eared seals; and Odobenidae, the walrus. Of the 19 true seal species, five are found in Antarctica. The Crabeater, Weddell, Ross, and Leopard seals each have their own genus, while the fifth, the Southern elephant seal, shares its genus with the Northern elephant seal (*Mirounga angustirostris*). The eared seals include sea lions and fur seals. There are five sea lion genera, and each has a single species. Named for the long mane of fur covering the beefy necks of the males, sea lions have a broad distribution, generally preferring temperate coasts but occasionally occupying equatorial niches or sub-polar zones. The nine fur seal species, distinguishable from sea lions by a thick waterproof pelt under long guard hairs, form two groups: the Northern fur seal has its own genus, while the eight southern fur seals form a second genus, *Arctocephalus*. Eared seals are less polar in their habits than many true seals, but most southern fur seals live in, or occasionally reach, sub-Antarctic or Antarctic regions. The Walrus, with its great tusks, is unique among seals. An exclusively Arctic animal, it is most closely related to the eared seals.

Seal physiognomy

Seals are well adapted for ocean life. The true seals have a thick layer of blubber for insulation in water, whereas the fur seals rely on their dense fur for insulation. The hind limbs of all seals are long, thin, and flattened, rather like paddles. Their faces are usually blunt, with reduced (in fur seals) or non-existent (in true seals) external ears. Their bodies are smooth and sinuous, and they swim by using

◀ LARGE MOUTH, LARGE PREY

All seals are carnivores, and the size of the mouth of each species is related to the size of its preferred prey. Adult Leopard seals usually feed on a range of prey that includes larger fish, penguins, and seals.

Seal family tree

Little is certain about the origins of seals. Many scientists place them in their own taxonomic order, Pinnipedia, while other authorities consider them a suborder of Carnivora. Some scientists believe that the closest living relatives to all seals are the weasels, badgers, skunks, and otters, while both modern and fossil cranial material suggests that all seals are descended from a bear-like ancestor.

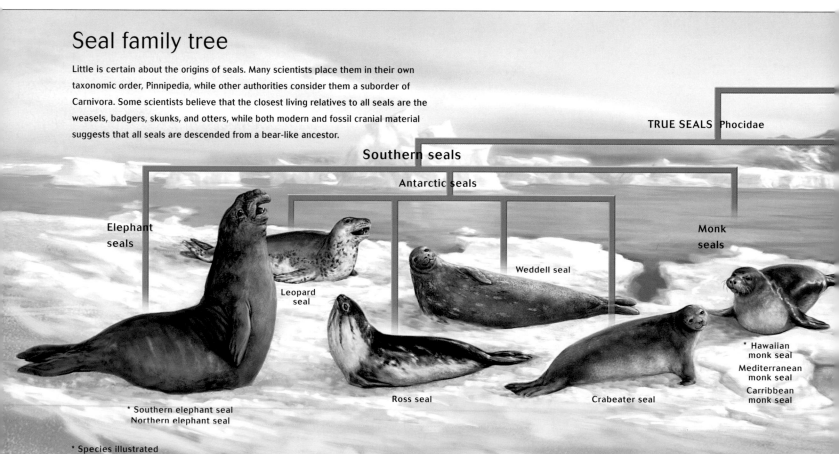

TRUE SEALS Phocidae

Southern seals

Antarctic seals

Elephant seals

Monk seals

Leopard seal

Weddell seal

* Hawaiian monk seal
Mediterranean monk seal
Carribbean monk seal

Ross seal

Crabeater seal

* Southern elephant seal
Northern elephant seal

* Species illustrated

their flippers and undulating their bodies. Most have a thick, heavy neck, but the vertebrae are only loosely interlocked, making them extremely flexible for a mammal. Agile and very strong, seals can withstand the impact of waves and currents and maneuver around ice and rocky shores.

Seals have slitted nostrils that close under water—another adaptation to marine life. When the nostrils are in the relaxed position they are closed, and water pressure increases the force keeping them closed. This means that when resting in water a seal does not need to concentrate on keeping its nostrils closed. In order to breathe, the seal must actively contract a pair of muscles. While at sea, seals can hold their breath for extended periods—elephant seals, for example, have been recorded holding their breath for two hours. When a seal does open its mouth under water, the tongue and soft palate close off the back of the throat, which means that the seal can take a mouthful of food without swallowing seawater. In addition, the larynx presses against the epiglottis to stop food and water from entering the lungs.

Eared seals have powerful muscles in their shoulders to support their strong front flippers, which they use as paddles in water and as feet on land. They can also bring their hind flippers forward and use them like legs to run across land. True seals, however, have lost this facility and when swimming, they use their hind limbs to undulate their hindquarters, where their main muscles are.

▲ STILL TIED TO THE LAND

As is true of all true and eared seals, this fur seal has a flexible, streamlined body with reduced peripheral appendages, which makes it well adapted to life in the water. But seals, unlike whales, are still connected to land. While a few seal species appear to mate in the water, most mate on land, and all come ashore (or onto ice) to give birth.

PINNIPEDS Pinnipedia

WALRUS Odobenidae

EARED SEALS Otariidae

Northern seals

Sea lions

Northern sea lion　Australian sea lion　*Hooker's sea lion　South American sea lion　California sea lion

Northern fur seal　Southern fur seals

Hooded seal　Gray seal　White-coated seals

* Common seal
Caspian seal
Ribbon seal
Harp seal
Ringed seal
Spotted seal
Baikal seal

Bearded seal

Walrus

* Antarctic fur seal
South American fur seal
New Zealand fur seal
Galapagos fur seal
Juan Fernandez fur seal
Australian fur seal
Guadalupe fur seal
Subantarctic fur seal

◀ FAT BUT WARM

Elephant seals use blubber as their primary insulation. It may look awkward on land, but it does not interfere with swimming. These seals often cover 70–80 kilometers (44–50 miles) per day when traveling to feeding locations.

This gives them a highly efficient swimming motion; however, their front flippers do not have the strength or the reach to be effective on land, and their rear flippers cannot bend forward in the way that the flippers of eared seals can. True seals do not run, but undulate across land, humping their body up with their flippers and throwing it forward with surprising speed.

The eyes have it

Seals' eyes have several adaptations for seeing under water, including a flattened cornea and pupils that can open very wide to admit as much light as possible. In land-dwelling animals the curved surface of the eye refracts light to project an image on to the retina in the back of the eyeball and the lens is used only for focusing, but in water the lens directs the image to the back of the eyeball. All seals have large, well developed eyes, which gather light more effectively than small ones. Elephant seals and Ross seals, both deep divers, have very large eyes with large lenses and can see well in deep water. Seals have retinas like those of terrestrial carnivores but lack cones, the sensory cells necessary to detect color. Their eyes consist entirely of rods—sensory cells that function well in low light. Elephant seal eyes have a modified photopigment like that of deep sea fish, perhaps to discern the bioluminescence of deep sea squid, their prey.

Out for the count

Antarctica poses unique problems in estimating population sizes. If you count those on land, what about seals in the water? How do you count seals that are hauled out on sea ice far away? The solution is to survey long narrow strips of sea from a ship or an airplane and count the seals hauled out on ice. Seals prefer to haul out at certain times of day. The time is noted and the proportion of seals presumed to be on the ice is factored into the estimate. The ice type and cover are also taken into account, as seal densities can be related to the area of pack ice available for them to haul out onto. Final estimates are likely to be low, as the proportion of seals actually on the ice will never be 100 percent; the best available estimates place the Crabeater seal population at about 14 million, but change the assumptions and the same counts suggest that the number may easily be as high as 40 million.

Eating for energy

Seals are active hunters, and as a group prey on everything from plankton to higher vertebrates. However, each species occupies its own ecological niche, and therefore has a tendency to exploit a particular type of prey. For example, Crabeater seals eat primarily krill, whereas Ross seals inhabiting the same regions will take mostly squid and fish.

Seal milk is a very high-energy drink, with a fat component of 30 to 60 percent and a protein content of at least 5 to 15 percent. (In comparison, cow's milk contains 2 to 4 percent fat and 1 to 3 percent protein.)

Some species—for example, some of the southern fur seals, many southern sea lions, and elephant seals—ingest rocks and sand. Rocks the size of a small orange have been recovered from the stomachs of seals, and one Southern elephant seal had 35 kilograms (80 lb) of rocks in his gut. These rocks, which are known as gastroliths, are found in such quantities that they are probably ingested intentionally, rather than by accident, and may provide stabilization while diving, or ballast to minimize the energy required for deep dives, or they may be for breaking up food or grinding up parasitic worms. Breeding and molting seals are more likely to have gastroliths, possibly because they need to fill their stomachs while they cannot feed at sea.

Mating and gestating

A female becomes pregnant immediately after mating, but the embryo stops growing after a few days and does not implant in the uterus for several months. It then develops in the normal way, and at the normal speed. Delayed implantation means that the pup is born almost a year after mating, although the period of fetal growth is much less than that, generally eight to nine months. Thus, birth and the next mating occur around the same time, so that the pup is born when it has the best chance of survival and the female has the most protected and shortest possible breeding and pupping period.

Females occasionally adopt orphaned pups if their own die. As most seals that lose pups are young mothers, this behavior may help them by providing much needed experience. Some pups find a foster mother after a normal period of nursing by their blood mother, and so are suckled for much longer than normal and gain many advantages from the resultant extra size and energy stores. Most foster mothers adopt a single pup, usually only when they lose their own pup when it is very young. In a few seals, however, the mothering instinct is so strong that they will nurse so many orphans they cannot produce enough milk to feed them properly.

> ICE FISHING
A Weddell seal with a large Antarctic toothfish (*Dissotichus mawsoni*) which it caught in 300-meter (980-ft) deep water under this hole in the fast ice near Cape Armitage, Ross Island. Fish are the main food of Weddell seals, but they also catch squid, bottom-living prawns, and octopus.

⌃ BACHELOR RIGHTS
A young bachelor Southern elephant seal rests on top of a weaner. Sometimes seals of different ages and sexes come ashore to rest and molt at the same time. A peaceful scene like this one may erupt into a short-lived aggressive outburst at any time. When caught in the fray, a smaller individual often becomes the victim of a larger seal's temper tantrum.

WATERPROOF JACKET

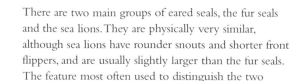

▼ WATERPROOF JACKET
Waterproof and nearly
windproof, a fur seal's fur is
made up of two layers—an
outer layer of coarse guard
hairs and a velvety underfur,
some 2–3 centimeters (about
1 in) thick.

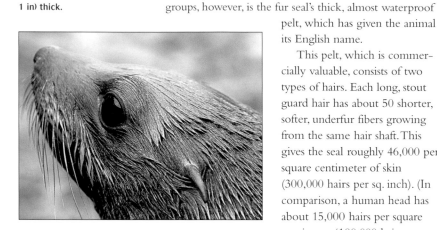

▼ AWAITING LUNCH
Two New Zealand fur seal pups
peer out from their hideaway
above the high tide mark.
After a mother has suckled
her pup for 10 days, she
mates—usually with the bull
controlling the territory. She
then leaves on her first feeding
trip, returning a few days later
to suckle her pup again.

Eared seals

There are two main groups of eared seals, the fur seals
and the sea lions. They are physically very similar,
although sea lions have rounder snouts and shorter front
flippers, and are usually slightly larger than the fur seals.
The feature most often used to distinguish the two
groups, however, is the fur seal's thick, almost waterproof
pelt, which has given the animal
its English name.

This pelt, which is commer-
cially valuable, consists of two
types of hairs. Each long, stout
guard hair has about 50 shorter,
softer, underfur fibers growing
from the same hair shaft. This
gives the seal roughly 46,000 per
square centimeter of skin
(300,000 hairs per sq. inch). (In
comparison, a human head has
about 15,000 hairs per square
centimeter/100,000 hairs per sq.
inch.) Sebaceous glands in the seal's skin produce enough
oil to make the fine hairs into a waterproof barrier that
traps air, so that only the tops of the guard hairs get wet.
In addition, the strong guard hairs serve to support the
fine underfur. In this manner, the fur seal's pelt is custom-
made to provide insulation when the animal is diving. Sea
lions, on the other hand, lack this underfur, and historically their pelts have had little financial value.

Fur seals and sea lions probably diverged into unique
groups only about two million years ago—a relatively
short time in evolutionary terms—and the two groups

still retain many similarities. Their social and reproductive
behaviors, for example, are very similar, and natural
hybrids have been observed where their populations
overlap in the wild.

On average, eared seals live for about 18 years. Males
probably live shorter lives than females because they must
vigorously compete with other males for breeding rights.
It is a stressful process for the males, and many do not
survive past early reproductive age.

The males of both the fur seals and the sea lions are
always much larger and heavier than the females. The
males have particularly solid-looking chests that are much
bigger than those of the females and are covered with
longer fur to increase their apparent size. The males use
these big chests for displaying to rival males in order to
intimidate them. The robust chest also offers protection
on the rare occasion that the displays degenerate into
active fighting between competing males.

Living spaces
The Otariidae occur from tropical, through temperate, to
polar environments, but are most commonly found in
temperate or sub-polar conditions. Sub-polar
populations are usually very large. These environments
tend to be more productive with nutrient-rich waters
that encourage large populations of prey species, which
eared seals can exploit on a seasonal basis.

All seals need to replace their hair, or fur, in a process
know as molting. This is particularly important for fur
seals, which rely on the waterproofing in their soft underfur for temperature regulation. Fur seals and sea lions
molt very gradually so that they can continue to swim
throughout this process and remain insulated.

> PLAYTIME
Eared seals can be playful, often surfing, tossing and dragging rocks, sticks, or other animals, chasing each other, wrestling, rolling over, waving their flippers, or biting their own tails. The purpose seems to be the formation of social bonds, and possibly the development of motor skills.

Claiming territories

Eared seals come together in breeding colonies on favored beaches to give birth and mate. The males will generally arrive before the females and spend the first few days fighting to establish territories, and they then maintain rank by vocalizing, posturing, and threatening. These breeding territories are often defined by natural boundaries, such as rocks or crevices, which allow easy identification of interlopers and provide clear-cut zones for posturing. In many species, the males return to the same area for as long as they can continue to win their place—usually only two or three years.

Cherchez la femme

Breeding fur seal males typically defend a specific patch of beach—taking a gamble that it will fill with females. The number of pups that a male fathers depends on the quality of his territory and the length of his tenure.

Females influence their breeding success by choosing the territory they use, sometimes moving through the territories of several males before settling. Males of many species try to herd females, usually without a great deal of luck because females will return to the territory of their choice.

Immediately after birth, the mother and her pup bond by nuzzling and vocalizing. When a female goes to sea to feed for the first time after giving birth, the pup stays close to where it suckled and the mother returns to the same area and locates her pup by sound and smell. When she calls, all the nearby pups respond, but she accepts only her own offspring. It is critical that the pup responds, because if it does not call the mother bites it or tosses it away, a behavior that has developed to stop her from feeding greedy or orphaned pups and thereby lessening the energy she has to nurture her own pup.

▲ FLUSHING THE SYSTEM
Like all mammals seals need water—yet many do not seem to drink it. Some seals get enough water from eating fishes that contain very little salt. The many species that live on invertebrates consume salty food, but seal kidneys are so effective that they can excrete salt in extremely concentrated urine and even extract fresh water from the seawater. Male eared seals are especially likely to drink seawater during their long breeding fast. Although they can derive some water through the oxidation of stored fat, they must replace water lost through panting, urination, and sweating. Presumably the net gain in fresh water from drinking seawater, while very small, is enough to be worthwhile.

FUR SEALS

Of the eight southern fur seals, genus *Arctocephalus*, five either permanently occupy or regularly visit sub-polar or polar regions.

South American fur seal

ARCTOCEPHALUS AUSTRALIS

OTHER NAMES: Southern fur seal, Falkland fur seal
LENGTH: female 1.4 m/4½ ft; male 1.9 m/6 ft
WEIGHT: female 50 kg/110 lb; male 160 kg/350 lb
STATUS: up to 750,000, sensitive to El Niño events
    IUCN: Lower Risk—least concern   CITES: not listed

South American fur seals range up the coasts of South America from Lima to Rio de Janeiro, throughout Tierra del Fuego, and to the Falkland Islands. They breed on islands and remote coasts, and congregate in small groups with the females well spaced. Pups are born in spring and early summer (mid-October to December), the peak time being November and early December. Mating occurs six to eight days after the female gives birth. Females nurse for at least eight months, and sometimes up to two years—a very long period for a fur seal. Most of the South American fur seal population is non-migratory, with some limited dispersal in winter, but most females remain near the colonies all year.

New Zealand fur seal

ARCTOCEPHALUS FORSTERI

OTHER NAME: Western Australian fur seal
LENGTH: female to 1.5 m/5 ft; male to 2.5 m/8 ft
WEIGHT: female to 70 kg/155 lb; male to 185 kg/410 lb
STATUS: 85,000–135,000; increasing at most sites
IUCN: not listed   CITES: Appendix II

New Zealand fur seals are found in New Zealand and nearby sub-Antarctic islands, in South Australia and Western Australia, and at Macquarie Island. These seals breed on islands and remote coasts in small to very large groups, with the females being well spaced. The males have massive necks, and thick manes that are used in dis-

play threats when establishing territories. New Zealand fur seals are non-migratory, with some dispersal after breeding and a return just before breeding, but most females stay near the colonies all year. They eat squid, octopus, fish, and even seabirds.

Females pup in late spring and summer (mid-November to mid-January), with births peaking in the second half of December. Mating occurs eight to nine days after birth, and peaks in late December to early January. Pups are weaned at about twelve months.

Antarctic fur seal

ARCTOCEPHALUS GAZELLA

OTHER NAME: Kerguelen fur seal
LENGTH: female 1.3 m/4 ft; male 1.9 m/6 ft
WEIGHT: female 35 kg/80 lb; male 135 kg/300 lb
STATUS: at least 2–4 million
    IUCN: Lower Risk—least concern   CITES: Appendix II

This colonial species breeds on islands south of the Polar Front, including the South Shetland Islands, South Georgia, the South Orkney and South Sandwich islands, Bouvetøya, and Heard and McDonald Islands, but is also found north of the Polar Front, on Marion and Prince Edward islands, Iles Kerguelen, and Macquarie Island.

Both males and females leave breeding colonies in winter. On land, they sometimes travel inland to lie on top of tussock grass behind the breeding beaches. They can move at 20 kilometers per hour (12 mph) on beach flats, and can outmaneuver humans through tussock grass. They make long trips during the breeding season, in which they eat krill, cephalopods, and fish. While at sea feeding females make more than 400 dives of between 5 and 100 meters (15 and 330 ft) in five to six days.

▼ **A BLONDE IN THE MIDDLE**
While most Antarctic fur seals have medium to dark brown fur, a small number are natural-born blondes. Pale, almost white fur seals are highly visible on dark rock and stand out in the crowd, but they don't seem to suffer any more difficulties catching food than their more camouflaged relatives.

South American and new zealand fur seals

0°
Tristan da Cunha
Gough I
South Georgia
Bouvetøya
Prince Edward Is
Sth Sandwich Is
Is Crozet
Falkland Is
Sth Orkney Is
Is Kerguelen
Heard I
90°W
South Pole
90°E
Balleny Is
Polar Front
Campbell I
Macquarie I
Antipodes Is
Auckland Is
Bounty Is
Chatham Is
100°

South American fur seal
New Zealand fur seal

> DISTINCTIVE FEATURES

The eared seals are generally smaller than the true or hair seals, and they are all very lithe and supple. The small external ear that characterizes the group can be seen on this Antarctic fur seal, as can the thick coat that made this species a popular target for sealers.

Antarctic fur seals give birth in late spring to early summer (late November to December), peaking in early December. Mating occurs six to seven days after the birth of the pup, and peaks in mid-December. Females nurse for four months or less, and the pups seem to initiate weaning. Pups are extremely active and begin practicing swimming in January, but do not become skilled swimmers until March.

South African fur seal

ARCTOCEPHALUS PUSILLUS

OTHER NAMES: Cape fur seal, Australian fur seal, Tasmanian fur seal, Victorian fur seal

LENGTH: female 1.8 m/6 ft; male to 2.4 m/8 ft

WEIGHT: female to 120 kg/270 lb; male to 360 kg/800 lb

STATUS: to 2 million in southern Africa, 50,000 in Australia

IUCN: Lower Risk—least concern   CITES: not listed

The South African fur seal ranges over southwestern Africa from Angola to Algoa Bay, and to southeastern Australia and Tasmania. It breeds on islands in medium to large colonies, with females relatively densely packed. It is polygynous, with female group sizes ranging from five to more than fifty. The species is non-migratory but shows some seasonal movement and some long-distance dispersal in winter. Individuals have been found

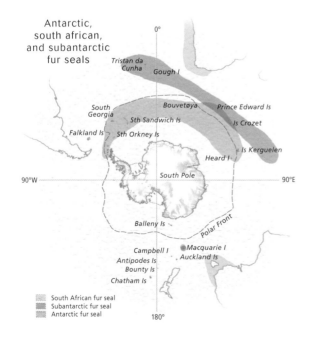

Antarctic, south african, and subantarctic fur seals

Tristan da Cunha
Gough I
South Georgia
Bouvetøya
Prince Edward Is
Sth Sandwich Is
Is Crozet
Falkland Is
Sth Orkney Is
Is Kerguelen
Heard I
South Pole
Balleny Is
Polar Front
Campbell I
Macquarie I
Antipodes Is
Auckland Is
Bounty Is
Chatham Is

0°
90°W
90°E
180°

- South African fur seal
- Subantarctic fur seal
- Antarctic fur seal

1,500 kilometers (950 miles) from the nearest colony. Feeding females generally stay at sea for approximately five days before returning to the colony. Unusually for an *Arctocephalus* species, mature bulls will tolerate the presence of young males more than one year old.

Pupping takes place in late spring and early summer (November and December), peaking in late November and early December. The peak mating period follows five to six days after birthing. South African fur seals generally nurse their offspring for 8 to 10 months.

Subantarctic fur seal

ARCTOCEPHALUS TROPICALIS

OTHER NAME: Amsterdam Island fur seal

LENGTH: female to 1.5 m/5 ft; male to 1.8 m/6 ft

WEIGHT: female to 50 kg/110 lb; male to 150 kg/330 lb

STATUS: slightly more than 310,000

IUCN: Lower Risk—least concern   CITES: Appendix II

This species breeds on islands and island groups north of the Polar Front, including Tristan da Cunha, Gough and Marion islands, Iles Crozet, Iles Amsterdam, and Macquarie Island. Breeding colonies vary in size and tend to be well spaced. Males move out to sea after breeding, while females and pups remain in the colonies. Their diet ranges from crested penguins and squid at Iles Amsterdam to cephalopods, fish, and krill at Marion Island. Females give birth in late spring to summer (late November to January), peaking in mid-December, and the peak mating period occurs 8 to 12 days after the birth. Pups are normally nursed for 9 to 11 months. Non-territorial males are not tolerated in the breeding colonies and form bachelor colonies with immature seals, spending 94 percent of their time completely inactive. Breeding males actively herd females and defend territories.

▲ AGILE ON LAND

Fur seals are extremely flexible, and can bring their hind limbs forward underneath their bodies and rise up on their front flippers. When moving fast, they usually bound forward with both hind limbs together, a motion well suited to moving around in their rocky habitats.

True seals

True seals, sometimes called hair seals to distinguish them from fur seals, belong to the family Phocidae. There are five Antarctic species of true seal: the Weddell seal, Leopard seal, Crabeater seal, Ross seal, and Southern elephant seal. All Antarctic true seals, together with the northern-hemisphere Monk seals and Northern elephant seal, are grouped together and called Monacine seals.

Streamlined, true seals are well designed for ocean life, and some are truly ice-dwelling. To cope with the extreme conditions in which they live, they have well developed layers of blubber instead of the thick fur that insulates fur seals. Even the thickest fur would be inadequate to meet the challenges of truly polar conditions.

True seals are more at home in water than they are on the land. They swim using their strongly muscled hind limbs, applying this power in a lateral stroke with their toes expanded. Their entire hind end undulates laterally, and they then straighten out their body in a recovery stroke, with the toes bunched together to minimize drag from the water. This is an extremely powerful stroke, which makes the true seals very capable swimmers.

True seals, however, are not as competent on land as they are in water because they have lost the ability to bring their rear flippers forward and use them as legs, in the way that eared seals do. True seals lumber along on the ground, using their whole body for movement, and their front flippers as levers. Those that live on ice have longer, stronger claws than those that live on land, and these long claws are extremely effective for clambering up on ice and moving around on ice ledges.

Mothering behavior

True seals have a very short period of pup dependence: as short as four days or as long as several months depending on the species. The main task of the female during this time is the efficient production and transfer of milk from mother to pup. However, their milk is highly nourishing,

WEDDELL SEAL

Tristan da Cunha
Gough I
Bouvetøya
Prince Edward Is
South Georgia
Sth Sandwich Is
Is Crozet
Falkland Is
Sth Orkney Is
Is Kerguelen
Heard I
South Pole
Balleny Is
Polar Front
Campbell I
Macquarie I
Antipodes Is
Auckland Is
Bounty Is
Chatham Is

Distribution
Vagrants

▼ A MOTHER'S FAST
A Weddell seal pup weighs 25–30 kilograms (55–66 lb) at birth and when weaned, at six or seven weeks, is about 100 kilograms (220 lb). While the mother nurses, she does not feed and drops from about 450 kilograms (990 lb), to about 300 kilograms (660 lb).

and this accelerates the nurturing process. Suckling pups can gain as much as 25 percent of their birth weight each day. Most true seal mothers fast while nursing, alternately resting and feeding the pup several times a day, an unusual pattern for mammals. They wean their young abruptly, simply by going to sea, leaving the pup to fend for itself.

Diving adaptations

Seals have 15 pairs of ribs, rather than the usual 13 pairs for mammals. This allows extra space for their slightly larger lungs. When a seal dives the pressure of other organs collapses its diaphragm against its lungs, forcing any remaining air out of its lungs. This protects the seal from any risk of the bends. Seals do hyperventilate before diving, but they store the oxygen in their blood and muscles and then expel the air. Seals have more blood than similar-sized terrestrial mammals. Seal blood also carries a higher proportion of hemoglobin, which carries the oxygen. So seals can carry about 3.5 times more oxygen per unit of body weight than humans can.

Weddell seal

LEPTONYCHOTES WEDDELLI
LENGTH: female 3.3 m/11 ft; male 3 m/10 ft
WEIGHT: 400–500 kg/900–1,125 lb
STATUS: approximately 800,000
IUCN: Lower Risk—least concern   CITES: not listed

This species is the most southerly breeding of all mammals. They tend to inhabit areas of coastal fast ice (ice that is connected to land) rather than the moving pack ice. Although Weddell seals favor the coastal fast ice areas of the Antarctic Continent and the nearby islands, they are occasionally found in South America, Australia, and New Zealand.

Particularly during winter, these seals breathe and enter the water through holes in the ice. They need to make or maintain these holes themselves, cutting the ice with their incisor teeth and enlarging and maintaining the openings with their canine teeth.

During the breeding season, females gather in small, loosely knit groups of as many as 100 individuals to give birth in the stable fast ice around the Southern Continent. They share the breathing holes that occur along cracks in the ice, while breeding males wait in the water beneath these entry points. Males defend an aquatic territory against other males, and by defending an entry hole, a single male can monopolize a large number of females.

The peak pupping period is from September through to November, but pupping may take place at any time from late August onward. The lactation period lasts for approximately 50 days, much longer than for most other phocids. This means that lactating Weddell seals must feed during the nursing period, unlike the other species of Antarctic true seals. Males that have won a territory wait for the females to wean their pups, and mate with them in November and December, when they enter the water.

Weddell seals are believed to consume a diet of fish, cephalopods, krill, and other invertebrates. They locate and catch this prey using a range of diving behaviors.

Baby fur

Most true seal pups have no insulating blubber, but they do have very fine, thick baby fur, called lanugo, that is very long in comparison with that of the adult seals. At birth the baby seal has virtually all the hair follicles that it will have as an adult, so these are much closer together than they will be when the animal has grown to full size. This combination of features means that the pup has a more effective insulating coat of fur than have its parents, and it very seldom gets cold. If a pup does begin to get cold, it can produce additional heat by increasing its metabolic rate.

➤ **RESTING BETWEEN DIVES**
If Weddell seals keep their dives short, they can dive all day, needing only a few minutes breathing air at the surface between dives. After a long dive (25 to 80 minutes) the Weddell seal needs more time to recover, sometimes well over an hour.

Wonderful whiskers

Whiskers, known scientifically as vibrissae, are very important to seals. They are prominent in most species, visibly clustered on the upper lip and above the eyes, with a single pair on top of the nose. There are many nerve endings at the base of whiskers, and they are extremely sensitive to motion. In marine mammals, vibrissae can pick up minute changes in the flow of water—even the minor disturbance caused by a fish's tail while it is swimming may be detected through a seal's whiskers. The vibrissae are pushed forward when a seal is chasing prey, possibly making them more effective at detecting fine differences in motion. This may enable the seal to predict the exact location of a fish by changes in motion rather than by sight, which is often of limited use especially at greater depths. Experimental trimming of the vibrissae of wild seals has shown that seals without their whiskers find it much harder to catch fish in the dark than do seals with their whiskers intact.

Leopard seal

HYDRURGA LEPTONYX

OTHER NAME: Sea leopard

LENGTH: female 3 m/10 ft; male 2.8 m/9 ft

WEIGHT: female 370 kg/815 lb; male 325 kg/715 lb

STATUS: 220,000 to 440,000

IUCN: Lower Risk—least concern   CITES: not listed

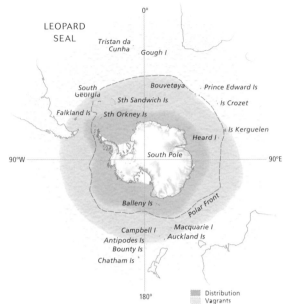

The Leopard seal is the largest of the five true seals in Antarctica after the Southern elephant seal. It is a true hunter, possessed of powerful jaws and a long, sinuous neck that it uses to great effect, retreating backwards then striking like a snake. This powerful animal can leap onto ice 2 meters (7 ft) above water level—an exit speed from water of about 6 meters (20 ft) a second. Leopard seals would rather be on ice than on land.

The Leopard seal's main habitat is the Antarctic pack ice, and its icebergs and smaller ice floes. Mature Leopard seals prefer to remain in the Antarctic pack ice, while youngsters can be observed throughout the sub-Antarctic zone from June to October each year. Adults from three to nine years old utilize the Antarctic coast in summer.

Leopard seals are known for their hunting prowess, even though over 65 percent of their diet usually consists of krill, fish, and squid. They take Crabeater seal pups and penguins. They usually capture adult penguins in the water, rarely hunting on land. There is one amazing record of four Leopard seals eating 15,000 Adélie penguins in 15 weeks; one animal had 16 adult penguins in its stomach.

▼ **WHERE ARE YOU GOING?** Leopard seals are closely related to Crabeater and Weddell seals, but have a sleek reptilian grace that their relatives lack. They do not form large breeding colonies, and their seasonal movements are not well known—although some individuals appear to disperse northward during winter, and young animals seem more likely to move about than adults.

Crabeater seal

LOBODON CARCINOPHAGUS

LENGTH: 2.35 m/7 ½ ft

WEIGHT: 220 kg/500 lb

STATUS: estimates range from 10 to 40 million, but probably about 15 million

IUCN: Lower Risk—least concern   CITES: not listed

These truly Antarctic seals populate the outer fringes of the pack ice, ranging among the icebergs and smaller floes, and recent evidence suggests they may like the deep

ice close to the Continent. Crabeaters do not truly migrate, but they do show some seasonal movement with the annual expansion and retreat of the ice.

Despite their common name, Crabeater seals do not live on crabs: they are specialist feeders that live mainly on krill. They also eat a small proportion of fish and squid, and other invertebrates. Their teeth are highly adapted for their diet, each tooth basically the same shape—extremely convoluted with a few twisted gaps. The complex cusps are designed to strain krill from the water.

Ross seal

OMMATOPHOCA ROSSI

> OTHER NAMES: Singing seal, Big-eyed seal
> LENGTH: to 2.4 m/8 ft
> WEIGHT: 220 kg/500 lb
> STATUS: thought to be approximately 220,000
> IUCN: Lower Risk—least concern   CITES: not listed

Ross seals live deep in the Antarctic pack ice, and are patchily distributed right around the southern polar regions. The Ross seal is a true pack-ice species and only rarely hauls out on land. Scientific knowledge of this species is limited because its preferred habitat of dense pack ice is difficult for humans to easily reach in ships or from land. At least part of the Ross seal population appears to undertake some seasonal movement, presumably in response to movements of pack ice and fluctuations in food availability. Ross seals can reach considerable depths in pursuit of their favored prey—cephalopods, fish, krill, and bottom-dwelling invertebrates.

Southern elephant seal

MIROUNGA LEONINA

> OTHER NAME: Southern sea elephant
> LENGTH: female 2–3 m/6$\frac{1}{2}$–10 ft; male to 5 m/15 ft
> WEIGHT: female 400 kg/900 lb; male to 3,700 kg/8,000 lb
> STATUS: 750,000 in 1985; some populations are declining
> IUCN: Lower Risk—least concern   CITES: Appendix II

The Southern elephant seal is the largest species of seal. Both sexes are solid-looking animals with a thick layer of blubber, but the bulls are very bulky, have exceptionally thick necks and chests, and are over nine times larger than the females. Despite their bulk, both sexes are surprisingly quick and nimble. Around their necks, mature bulls usually show a great deal of heavy scarring from fighting—an awesome sight to behold—and cows may have small scars on the head and neck incurred during mating.

The male elephant seal has an inflatable nose, which is a sign of adulthood and extends fully only when the bulls are completely mature, at about eight years of age. At this time, the tip of the nose can hang down below the animal's mouth, and its nostrils point straight down. The nose has a massively enlarged nasal cavity, which a breeding male can fill with air. The whole nose can then be made erect by increasing the blood pressure to the area. At this point, a large, raised cushion along the ridge of the nose is formed. The enlarged cavity may act as a resonating chamber for the male's roaring, which he uses to intimidate other bull seals.

Southern elephant seals spend most of their lives at sea in Antarctic and sub-Antarctic oceans, where they eat cephalopods and fish. They breed on islands on both sides of the Polar Front, on the Argentinian coast in Patagonia, and on the Antarctic mainland—though rarely. Non-breeding animals are regularly spotted along the coasts of Antarctica. One long-distance traveler was tagged in South Georgia and some time later spotted in South Africa, a swim of about 4,800 kilometers (3,000 miles).   LW

Whales

Whales, dolphins, and porpoises belong to the Order Cetacea, from a Greek word meaning "large sea creature." And many are indeed large—no dinosaur ever reached the size of a Blue whale which, at 30 meters (100 feet) long, is the largest animal ever known to exist. Not all cetaceans are large, however; some dolphins reach only 1.2 meters (4 feet) but are still classified as whales. There are two sub-orders of whales: Mysticeti, the baleen whales, and Odontoceti, the toothed whales. Of the almost 90 known whale species, 11 are baleen feeders and 76 are toothed species, which are generally smaller than the baleen whales. Dolphins and porpoises are toothed whales.

All whales are carnivorous, but different species favor different types of food. In one of nature's paradoxes, the huge baleen feeders mainly live on zooplankton, filtering enormous quantities of these minute sea creatures from the water through comb-like plates in their mouths called baleen. Toothed whales use echolocation to hunt larger animals, such as fish and squid, and in Antarctic waters Orcas hunt warm-blooded prey such as penguins, seals, and even other cetaceans.

Whales inhabit all the world's seas, from shallow tropical waters to the deepest and coldest oceans. Their lives revolve around two major requirements—feeding, and breeding. The feeding season for most baleen whales is determined by the annual cycle of productivity in polar waters. During the summer feeding season, sea ice is at its minimum extent, and zooplankton including krill are most abundant and easily found. During winter, when ice covers the sea, krill are more difficult to find, and the majority of baleen whales head to warmer tropical or temperate waters to breed. The reason for this is still not certain, but it is thought likely that warm water allows calves to channel their energy into growth in early life, when their insulating blubber is thin. However, it is now known that some Minkes and Orcas, at least, remain in sea ice during winter, and may breed there. Because warm seas tend to be food-poor, most baleen whales have evolved a cycle of feast and famine, storing enough energy in blubber to fuel them during their winter migrations.

Living in water

Whales are mammals, and like all mammals they must breathe air. But because they have adapted over tens of millions of years to a totally aquatic existence, they differ from land mammals in many ways. Two of the important differences are how they breathe, and—a related problem—how they give birth in water.

Most mammals breathe through their mouth as well as their nostrils. As an adaptation to aquatic life, however, whales have an airway that is separate from their mouth, and their nostrils (the blowholes) are on the top of the

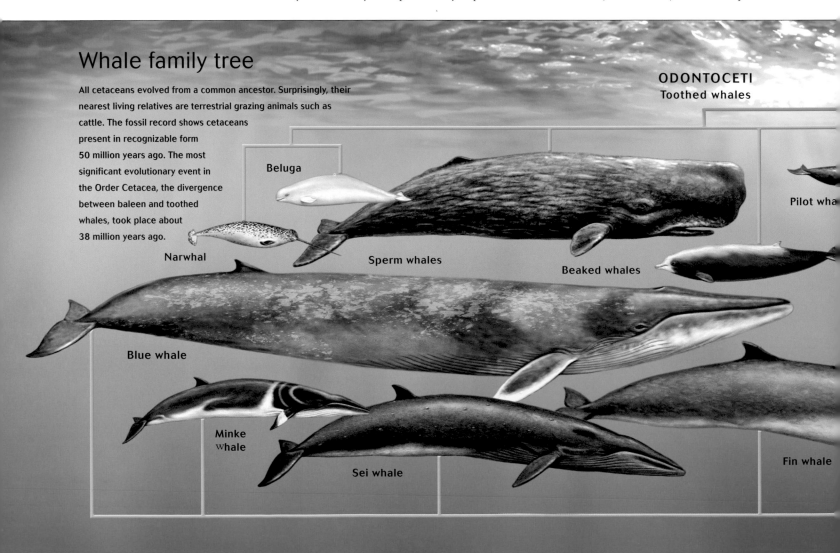

Whale family tree

All cetaceans evolved from a common ancestor. Surprisingly, their nearest living relatives are terrestrial grazing animals such as cattle. The fossil record shows cetaceans present in recognizable form 50 million years ago. The most significant evolutionary event in the Order Cetacea, the divergence between baleen and toothed whales, took place about 38 million years ago.

ODONTOCETI
Toothed whales

Beluga

Pilot wha

Narwhal

Sperm whales

Beaked whales

Blue whale

Minke whale

Sei whale

Fin whale

head for efficient breathing when surfacing. This also means that baleen whales can breathe while simultaneously filling their mouths with water and food; which may be particularly important for suckling calves. Unlike humans, whales are voluntary breathers: they hold their breath and choose when to breathe. The breath is often exhaled explosively—the blow—followed by a very rapid inhalation before diving again. Newborn calves are often helped to the surface for their first breath, assisted in some species by "aunts," who accompany the mother during birthing.

In other adaptations to an aquatic lifestyle, whales have virtually no hair, no sweat glands, no external ears, no external male sex organs, and no hind limbs. Although they do not have gills or scales as fish do, they have developed some fish-like characteristics—a streamlined, torpedo-like form, for example. Unlike fish (which have vertical fins and tails that move from side to side), whales have horizontal tail fins, called flukes, that move up and down. Cetaceans have evolved to move through water with minimal turbulence compared with most other mammals.

Temperature control

Being warm-blooded, whales need to maintain a constant body core temperature. Water quickly draws heat from warm bodies, and some species which migrate to polar waters have to contend with sea temperatures as low as

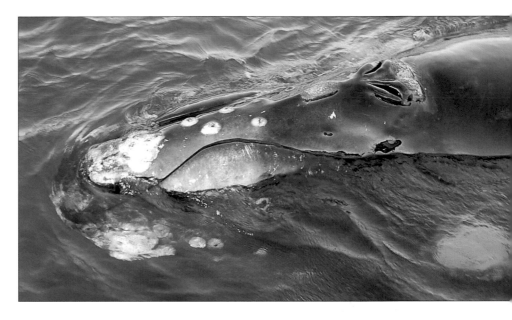

−2°C (28°F). To cope with this, they carry a warm coat of blubber, a mixture of fat, oil, and connective tissue, which lies under their skin. In some species, blubber may make up as much as 30 to 40 percent of total body weight, and is the energy store which fuels long winter migrations, including the production of calves and milk. Humpback whalers in New Zealand and elsewhere caught whales on their northward migration to breeding areas, while they were heavy with blubber, and ignored them on their southward migration at the end of winter, when

▲ SEASONAL MOVER
A Southern right whale near the Auckland Islands, a winter breeding area where mating and calving take place. In summer, Southern rights migrate southward to rich feeding grounds in the Southern Ocean.

CETACEA
Whales

MYSTICETI
Baleen whales

Dolphins

River dolphins

Porpoises

Orca

Gray whale

Right whales

Bowhead whale

Humpback whale

▲ SURVIVAL SKILLS
Insulating blubber enables the Minke whale to live throughout the year in ice-packed Antarctic waters. The smallest of the baleen whales, Minkes spend their lives locating krill swarms beneath the ice, while at the same time playing a cat-and-mouse game of survival with their own predators—Orcas. Minkes can hold their breath for only 20 minutes or so, and may use echoes of their own calls to detect vital breathing holes.

they were thin and hungry. Many species that migrate to tropical waters to give birth may do so for the benefit of their infants: the calves have time to grow a layer of blubber before encountering colder waters for the first time. On the other hand, some species (including Orcas, Bowheads and Narwhals) live their lives in icy waters and bear their calves there, and these calves are quite capable of surviving these conditions.

Whales also have a heat exchange mechanism in the circulatory system that helps to regulate their body temperature. In mammals generally, warm blood travels to the extremities and cooler blood, stripped of its oxygen, returns to the lungs to be replenished. The whales' blood system is different; they have inter-woven inward and outward blood vessels. When they are cold, they can remove the heat from the blood traveling to the extremities and bring it back to the core to keep their vital organs warm. Conversely, when a whale becomes too hot, it is able to direct warm blood to its body surface, especially the fins. The colder water absorbs the heat, and the cooled blood then returns to lower the temperature of the internal organs.

A consequence of large size in whales is that larger bodies conserve heat more efficiently because of the comparatively low ratio of skin surface to body mass. Animals with a larger body mass have proportionally less surface area to lose heat through than small animals. One possible cause of the evolution of large size is to enable whales to spend more time in cold, food-rich waters.

Cetacean anatomy
Whales moved into the oceans millions of years ago, and they have evolved adaptations that are not seen in land animals. For example, they have no gall bladder or appendix, as humans do. They also do not have carotid arteries; instead, they have small blood vessels, the retia mirabilia, that carry blood to the brain and may have a role in maintaining consistent pressure in the blood flowing to the brain, particularly during deep dives.

Some whales have become so large because living in water, they are not as subject to the constraints of gravity as terrestrial animals are. When large whales strand, they may be crushed under their own weight. The bones of whales are spongy and filled with oil, which probably increases their buoyancy, although most whales will sink when dead. The skeletons of whales are flexible in the vertical plane, but generally they lack sideways flexibility. Neck vertebrae fuse during maturation to support the head as it travels forward through the water. The head itself has become elongated during the course of evolution, to support the feeding structures such as baleen or an array of teeth. The front limbs have become flattened paddles, which are used for steering and control, yet their bone structure still shows their common mammalian ancestry in the form of five "fingers." Towards the tail, the vertebral column becomes more tapered and flexible, permitting the vertical movement used for locomotion. The flukes themselves are not supported by bone, but by muscle and firm, elastic connective tissue. Whales do not have any trace of leg bones, and the vestiges of the pelvic girdle (the hips) have dwindled to two small bones that are embedded in the body wall rather than attached to the skeletal structure. In land mammals these bones support the hind limbs, but in whales their only purpose is to support the muscles of the male's external reproductive organs, which are usually hidden in a slit that runs along the under-side of the body.

Taking to the water
Prehistoric cetaceans are among the earliest recorded marine mammals, and fossil cetaceans have been found on all continents. Aquatic dinosaurs died out 65 million years ago, leaving oceanic niches to fill. Mammals took to the water, and cetaceans were already present 50 million years ago. There are fossils of animals with webbed hind feet (semi-aquatic precursors of whales), and evidence that whales descended from a wolf-like, hoofed land mammal with large teeth, also thought to be an ancestor of even-toed hoofed mammals such as cattle. (An alternative but less likely theory holds that whales are more closely related to odd-toed hoofed mammals such as horses.) Cetaceans gradually became more efficient in water; by 38 million years ago, toothed and baleen whales had evolved their main characteristics, and 13 million years later different species had evolved within the two groups. About 15 million years ago, climatic cooling produced the nutrient-rich polar seas of today, and larger whales seem to have evolved as a result. Modern groups of whales had largely emerged by approximately seven million years ago.

Baleen and toothed whales

The main difference between baleen and toothed whales relates to feeding methods. Baleen whales (below) have evolved flexible baleen plates for filtering small schooling prey. Toothed whales (top) instead have sharp teeth for grasping larger individual prey items. In addition, baleen whales have retained two nostrils, while toothed whales have only one.

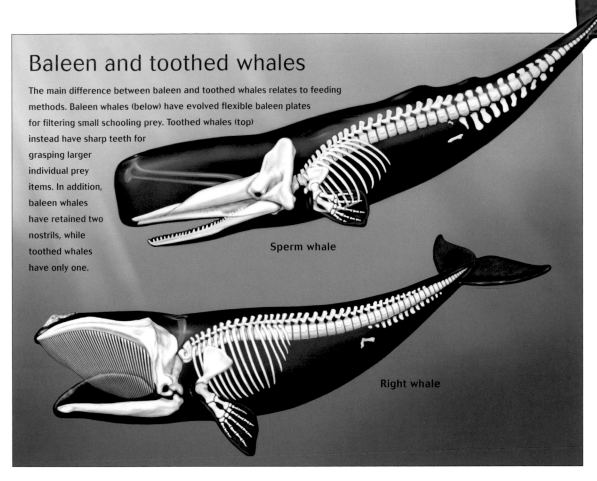

Sperm whale

Right whale

▼ TO DARK AND ICY DEPTHS With flukes raised, a Sperm whale begins a deep vertical dive to feed on squid. Sperm whales reach phenomenal depths—some 3,000 meters (9,850 feet) or more—and can remain submerged for more than two hours. However, most dives are much shallower and shorter. Whales use loud clicks both to communicate with each other and to echolocate their elusive, fast-moving prey.

Baleen whales

Only about 15 percent of whale species are baleen whales—but what they lack in numbers they make up for in size. Even the smallest baleen whale, the Pygmy right whale, reaches 6 meters (20 ft), and the mighty Blue whale grows to about 30 meters (100 ft).

Unlike the toothed whales, which have only one blowhole, baleen whales have two blowholes protected by a raised splashguard, or coaming, that prevents water from entering the blowholes while they are breathing.

Baleen whales are named for the comb-like baleen plates that hang from their upper jaws, which are used to strain their prey from seawater. The baleen whales are divided into two groups: rorquals (the groove-throated whales) and right whales. Right whales have enormously long baleen, and continuously strain prey as they swim, while most rorquals rapidly engulf huge mouthfuls of food, extending their throat capacity by ballooning out their throat pleats, creating a huge sac containing water and food. The muscular pleats are then contracted, squeezing water through the filtering baleen while their prey remains inside.

In order to catch enough food, these whales need enormous jaws that can provide enough room for the filtering baleen. This requirement has resulted in major modifications to the shape of the baleen whale skull, which is very different to that of the toothed whale. The head of some species, such as right whale, may be as much as a third of its body length. Each species of baleen whale has a different color, thickness, and number of baleen, depending on their feeding habits. Coarse baleen filters out larger prey, whereas finer baleen filters smaller prey. Right whales, which generally hunt very small prey, have long, very fine baleen as well as a bowed upper jaw to accommodate the length of baleen.

While baleen whales are known for feeding on zooplankton, they also devour small fish, particularly schooling species like herring, cod, mackerel, capelin, pilchards, and sardines. These species, or bottom fish like cod and sand lance—and even the occasional squid—make up the bulk of baleen whale food in areas, especially in the northern hemisphere, where small schooling fish are abundant. However, in the cold southern waters of Antarctica, zooplankton, particularly krill, is the whales' principal food source.

The distribution of krill—and of baleen whales—is determined by the dynamic oceanographic processes which influence phytoplankton production. These minute sea plants—and krill—live under and among sea ice during winter, and as the ice melts back during spring and summer, migratory baleen whales follow its retreat to feed. While it was long considered that the southern seas were largely empty during winter, recent studies have shown that Minke whales are present year-round, their small size possibly allowing them to utilize the scattered winter food supply.

Whale talk

All cetaceans produce sound for communication, and at least some for echolocation. Toothed whales, which are known to echolocate, generally use higher frequencies than do baleen whales, which are not yet proven to echolocate. Higher frequencies travel only short distances, while very low frequencies may travel hundreds or even thousands of kilometers. Some dolphins produce sounds which are too high-pitched for the human ear, while the deep rumbles of Fin and Blue whales are below our hearing range. In between, most species produce sounds that can be heard by the human ear. Many of us are familiar with the haunting songs of Humpback whales, or the whistles and clicks of bowriding dolphins.

▲ BREATHING EASY
A rorqual surfaces, showing the twin blowholes characteristic of baleen whales. Nostrils on the top of the head enable these whales to breathe while mostly submerged. Many fast-moving species also have a raised splashguard (seen here), so they can breathe while swimming, even in rough seas.

◄ PARASITES
Larger whales all collect various external parasites, including diatoms, whale lice, and barnacles. Whale lice congregate on the rough white patches (callosities) that are unique to right whales.

RIGHT WHALES

Right whales were so named by whalers because they were the "right" whale to hunt—they inhabited coastal waters, they were slow swimmers, they floated when dead, and they provided valuable baleen and high quality oil. All these factors contributed to the mass slaughter of right whales in the nineteenth and early twentieth centuries.

The presence of white patches (callosities) around the head is a positive identification of a right whale. Callosities are raised, horny layers of skin that may be heavily infested with barnacles, parasitic worms, and whale lice. They are present from birth, and males have more than females; they may act as weapons against other males.

The five right whale species—Bowhead, North Pacific right, North Atlantic right, Southern right and Pygmy right—occupy the earth's cooler seas. The Bowhead is restricted to Arctic waters, and the two northern rights to their respective ocean basins. The Southern right and Pygmy right whales share southern cool temperate waters, with Southern rights penetrating further south into polar waters.

Southern right whale

EUBALAENA AUSTRALIS

FAMILY: Balaenidae

LENGTH: female 14 m (45 ft); male 13.7 m (44½ ft)

WEIGHT: female 23 tonnes, male 22 tonnes

STATUS: approximately 7,000, increasing; IUCN: Lower Risk—conservation dependent   CITES: Appendix I

Southern right whales range from Antarctic to temperate waters. The species reaches the northern end of its distribution during winter, which they spend off the coasts of the southern parts of Africa, South America, New Zealand, Australia, and sub-Antarctic islands. During their breeding season in winter, Southern rights come very close to land, preferring shallow protected bays. They then travel south to feed in the plankton-rich sub-Antarctic waters during summer. Southern right whales have shorter migration routes than most other baleen whales, because they do not migrate to tropical waters. As their numbers are increasing, however, there is evidence that their migratory range is expanding into both warmer and colder seas—there have been increased sightings along the Antarctic sea ice edge in recent years. It is thought that they may now be re-occupying habitat they have avoided since the whaling era when their numbers were so drastically reduced. Right whales are comparatively slow swimmers, seldom exceeding 8 kilometers per hour (5 mph).

Right whales have an extensive repertoire of sounds, which include a loud bellow that is sounded when their blowholes are above water, and which can be heard several hundred meters away. Southern rights also produce low-frequency sounds, which have been recorded during travel, courtship, and play, but these are not prolonged and repetitive like Humpback songs. However, they do make grouped sounds called belches or moans, ranging from 50 Hz to 500 Hz, for up to a minute. The purpose of these different noises is not yet known, nor is it understood why Southern right whales appear to be most vocal at night.

▲ MOTHER AND CALF

A Southern right whale and her white calf. The calf is not an albino, but is probably a male with a recessive gene for pale skin (white female calves are not known). The color will darken to a milk-chocolate brown during its first year of life, but never to the brown-black of its mother.

RORQUALS

These are the most abundant and diverse group of baleen whales. Their name comes from the Norse word for a tubed or furrowed whale, which refers to the grooves, or pleats, that run the length of the throat and which allow it to expand when the whale is feeding.

There are six rorqual species. The Minke, Blue, Fin, Sei, and Bryde's whales form the genus *Balaenoptera*, while the Humpback has its own genus, *Megaptera*. The size difference among Balaenoptera is huge: an adult Minke, at 10 meters (33 ft), is only just larger than a newborn Blue whale, which may grow to 30 meters (100 ft). Most rorquals migrate seasonally between warm waters, where they breed and calve, and cool temperate to polar waters, where feeding occurs—but Bryde's whale is found only in tropical and warm temperate waters. The predictability of the timing and destinations of rorqual migrations made it easy for early whalers to locate and hunt the larger species.

▼ **TELLTALE BLOW**

The blow of the Blue whale is unmatched in height and power. Keen-eyed Antarctic whalers could spot these blows from a great distance, a factor that contributed greatly to the near extinction of Blue whales during the early twentieth century. Now researchers, not whalers, scan the waters, hoping to learn more about this magnificent species and so help protect it into an uncertain future.

▼ **WHY SO BLUE?**

Seen from space, the earth is blue. Seen underwater from above, a Blue whale is also blue. This is puzzling, as the true skin color of a Blue whale, when viewed above the water surface, may vary from silver to almost black. Yet once they slip beneath the waters there is an instant transformation, as if a light is switched on.

Minke

BALAENOPTERA BONAERENSIS

OTHER NAMES: Lesser rorqual, Piked whale, Pikehead
LENGTH: female 10 m/33 ft, male 8 m/26 ft
WEIGHT: up to 13 tonnes
STATUS: northern hemisphere: 60,000–80,000
southern hemisphere: 300,000–1 million
IUCN: Lower Risk—near threatened   CITES: Appendix I

Minkes are found in all of the world's oceans, and they are equally at home in the Arctic, the Antarctic, and the tropics. While many inhabit the open ocean, they are just as capable of living in the heaviest of Antarctic sea ice.

Growing to only about 10 meters (33 ft), the Minke whale is the smallest of all the rorquals. It is very sleek in shape and its head is sharply pointed. Its body is dark gray on top, graduating through various shades to a pale underside, and it often has a whitish patch on its flipper. Unlike the larger rorquals, who breed only every two or three years, there is some evidence to suggest that Minkes probably have an annual breeding cycle. Minkes mainly eat krill, but will also eat fish and the occasional mollusk.

Sei whale

BALAENOPTERA BOREALIS

OTHER NAMES: Pollack whale, Short-headed
   sperm, Japan finner, Sardine whale
LENGTH: female 17 m/56 ft; male 15 m/50 ft
WEIGHT: 20–25 tonnes
STATUS: unknown population, thought to be
   greatly reduced from original numbers
   IUCN: Endangered   CITES: Appendix I

Sei whales occupy cold, temperate, and tropical waters across the world. Seis are often found away from the coastal edges of the seas, in the open ocean, where they gather along oceanic fronts in pods of three or more.

Sei whales will devour almost anything they can find, but they seem to prefer smaller species such as copepods and amphipods. They will consume larger crustaceans such as krill, as well as fish and squid, but they are basically adapted to hunt smaller species. One of the favorite techniques of these whales is skimming, where they swim along the surface, twisting through the water and skimming zooplankton into their open mouths.

Fin whale

BALAENOPTERA PHYSALUS

OTHER NAMES: Common rorqual, Finner, finback,
Razorback
LENGTH: female 26m/85ft, male 21m/69ft
WEIGHT: up to 80 tonnes
STATUS: northern hemisphere population estimated at 40,000; southern hemishere:
5,000–30,000
IUCN: Endangered   CITES: Appendix I

> DRAWN BY CURIOSITY

Inquisitive Humpbacks investigate the photographer's boat in the placid waters of the Antarctic Peninsula. The whale on the left is spyhopping (raising its head above the surface to get a clearer view), while its companion prepares to dive under the boat. Humpbacks are among the most curious of whales; if they have no pressing business, they will often spend considerable time around and under vessels.

Fin whales are dark gray above, with swirling patterns known as chevrons on their backs, and white underneath. Although they are the largest whale after the Blue whale, they are slimmer, and not particularly heavy given their size. It is the only whale with a distinctive asymmetrical body coloration, with its right lower jaw being white, while the left side is dark. The baleen plates also follow this pattern. This coloration may be an advantage when feeding as it possibly helps to confuse their prey.

Fin whales range from polar to tropical waters. There are clearly discrete populations which are highly migratory, and many individuals have been identified returning to mate in their mother's breeding grounds. Fin whales are normally seen in pods of six to ten, but they are also often found in pairs. The size of the group may be related to the amount of food in the area, or to courtship and mating.

Blue whale

BALAENOPTERA MUSCULUS

OTHER NAME: Sulfurbottom

LENGTH: female 31 m/102 ft, male 30 m/100 ft

WEIGHT: 100–120 tonnes, possibly up to 150 tonnes

STATUS: unknown population, but probably less than 10,000

IUCN: Endangered   CITES: Appendix I

The Blue whale is probably the largest animal ever to have lived; *Brachiosaurus*, the biggest known dinosaur, reached a length of only 23 meters (75 ft), compared with the Blue whale's 30 meters (100 ft). The largest accurately measured Blue whales were both females. The longest had a body length of 33.6 meters (110 ft), and the heaviest weighed 190 tonnes and measured 27.6 meters (90 ft). Larger animals live longer than smaller ones, and the Blue whale is no exception; some individuals are thought to have lived for 110 years.

Although it is so large, the Blue whale is sleek and slender. Like most rorquals, it is streamlined and is a comparatively fast swimmer, having a torpedo-shaped body with no unnecessary appendages that could drag. In both males and females, a slit along the underbelly contains and protects the sexual organs when not in use.

The Blue whale is a cosmopolitan species, living in all oceans, and ranging from polar to equatorial waters. When in Antarctic waters, it will follow the ice edge seeking krill swarms (to satisfy the needs of its huge body, an adult Blue whale must catch a total weight of around 3.6 tonnes of krill each day). It may go deeper into the

Antarctic ice than all other baleen whales, except Minkes, and like most baleen whales migrates from the poles to the tropics at summer's end. Blue whales seem to gather in low numbers, which may be the effect of slow recovery from large-scale whaling.

Humpback whale

MEGAPTERA NOVAEANGLIAE

LENGTH: 15 m/50 ft

WEIGHT: generally 25-30 tonnes; up to 36 tonnes

STATUS: possibly 20,000–30,000

IUCN: Vulnerable   CITES: Appendix I

Humpback whales are much heavier in build and are slower swimmers than other rorquals, and so were heavily hunted by whalers. Though only 15 meters (50 ft) long, they can weigh up to 40 tonnes. A distinctive feature is their long pectoral fins that are about a third of body length. These fins make them very maneuverable and, despite their stocky appearance, they are graceful acrobats underwater.

While their coloration can vary from all black to all white, they tend to be dark above and white below. The undersides of their tail flukes, and often their flanks too, are pigmented with distinctive patterns which enable photo-identification of individuals. They hunch their back when they dive—hence their common name—and there are a series of lumps, or tubercles, around their head. They are one of the few whales to grow any hair.

Humpbacks are well known as the "singing whales." They sing complex, repetitive songs, each of which may last for 10 to 15 minutes. The individual sounds or "units" of Humpback song have been described as yaps, snorts, chirps, groans, whooos, eees, and ooos. Units are organized into "phrases," which, in turn, are organized into "themes"—and the themes are organized into songs. Songs may be repeated for hours, and have been heard at distances of at least 31 kilometers (20 miles).

⌃ SLEEK AND FAST

A Sei whale, recognizable by its tall, hooked dorsal fin, surfaces to breathe. Sei sometimes exploit copepod swarms in cool-temperate or sub-Antarctic waters; at other times they move south into icy waters to feed on krill. Little is known of their social organization, migrations, or breeding grounds.

Toothed whales

The best known of all whales must be Moby Dick, the great white Sperm whale immortalized in the eponymous novel by American writer Herman Melville. Moby Dick was a giant male Sperm whale, and it is this whale's tall, blunt profile that often features in children's books and drawings of whales. The Sperm whale is the largest of all the toothed whales and can reach 20 meters (65 ft) in length. However, its vast and impressive size and shape are unique among toothed whales. The dolphins, porpoises, and beaked whales that make up the bulk of toothed whale species are generally much smaller, some of them barely more than 1 meter (3 ft), and most of them under 5 meters (16 ft). They also range in habitat from steamy jungle rivers to ice-covered seas, and many of them live their lives rarely seen by humans.

Toothed whales hunt larger prey than do baleen whales, feeding on fish, mollusks, birds, and mammals, indeed, almost any animal they can catch—up to and including baleen and other toothed whales. Each toothed species has its preferred hunting patterns and prey: some home in on cephalopods such as octopus, squid, and cuttlefish; others specialize in grubbing up bottom-dwelling worms and crustaceans. Dolphins use their conical teeth to grab fish and swallow them whole, whereas porpoises have shearing teeth to rip prey apart before swallowing it.

Like baleen whales, toothed whales have a smooth, torpedo-like body but, due to their smaller size, they are far more flexible and agile than their more ponderous baleen cousins, and are far less likely to be encrusted with barnacles and whale lice. As well as using their teeth to grasp prey, toothed whales may use them for fighting, and sexually mature animals of both sexes are often savagely striped with scars inflicted by the teeth of other whales.

Head pieces

Most toothed whales have a melon—a large, lens-shaped pocket of fatty oil at the front of their head. The melon is more highly developed in species that feed in very deep water where light does not penetrate, or in muddy river water where visibility is poor. It is thought that the melon acts as a lens to focus echolocation sounds, which are generated inside the skull, into a beam of varying width and intensity. Returning echoes are received through oil-filled channels in the lower jaw. Echolocation is a sense that is dimly understood—humans have nothing comparable. We do know it is used for hunting prey, for navigation, and possibly for investigating the emotional state of companions, or what their last meal was.

Toothed whales have only one external blowhole. They once had two nostrils, but these have merged within the whale's head, and a single nasal passage reaches the skin.

All toothed whales have teeth at some stage, but these teeth are often vestigial or highly modified, particularly in the beaked whales which may have just one pair of teeth in the lower jaw. This raises the question of how these animals with minimal dentition are able to secure their prey. The answer is that they use a sudden, powerful suction once they are near enough to their prey for it to be effective. At the other end of the scale, those species with functional teeth may have more than 260 of them. These are usually simple, conical pegs, but some species have evolved highly specific tooth shapes that are of little practical use in normal feeding. Male Narwhals are the most dramatic example of this development. They have one normal tooth and one very long, spirally twisted tusk, and adult males have been seen using these to "joust" with each other. Many beaked whales have similar but less spectacular dental equipment.

Orca

ORCINUS ORCA

OTHER NAMES: Killer whale, Blackfish, Grampus
LENGTH: female 8.5 m/28 ft; male 10 m/32 ft
WEIGHT: female 7.5 tonnes; male 11 tonnes
DISTRIBUTION AND HABITAT: worldwide; found in all ocean habitats, from the tropics to the polar ice, but most common near rich food sources in high polar latitudes
STATUS: unknown population, but considered locally abundant
IUCN: Lower Risk—conservation dependent
CITES: Appendix II
APPEARANCE: These are the largest of the dolphins, and they are huge by dolphin standards; their size, and their black and white panda-like pattern, make them impossible to miss. Mature males are almost twice as large as adult females, and they have an exceptionally tall dorsal fin, and very large fins and flukes.

The Orca, or Killer whale, is one of the most successful dolphin species, with populations occurring in all the

world's seas and oceans. Orcas have 44 conical teeth with which they seize their prey. They are frequently referred to as Killer whales, which is somewhat undeserved because every other cetacean species also kills to eat. Furthermore, they have never been known to attack people in the wild, even when provoked. Unfortunately, the reverse is not true: Orcas are often killed because of their propensity for stealing from human fisheries. To such a hunter, taking fish from longline hooks is as easy as picking grapes.

An adult Orca needs to consume an average of 68 kilograms (150 lbs) of seafood each day. Orcas usually hunt in packs, and will show a remarkably disciplined division of labor, such as when attacking large whales. Some populations have discovered how to hunt seals resting on shore by surfing in and out on waves—this is probably the closest any modern whale gets to going on land. No matter what the target, Orcas are efficient, cooperative hunters. Depending on their location, they will eat whatever food is available, including fish, squid, sharks, stingrays, shellfish, seabirds, dugongs, and seals, as well as dolphins, porpoises, and baleen whales up to, and including, Blue whales. No predator is known to prey on Orcas, however.

Off the coast of British Columbia and Alaska, there are two different types of Orcas: the residents, which inhabit a home range and eat mostly fish; and the transients, which range much more widely, and favor marine mammals as prey. Recently a third, more flexible offshore type has been identified. The basis of Orca society is a

DETERMINED HUNTER

An Orca in its element—the Antarctic sea ice, where it hunts penguins, fish, seals, and whales. Orcas have been seen rushing at ice floes where seals are resting, creating a bow wave that rocks the floe and tips the hapless prey into the water. They are formidable predators capable of complex, coordinated attacks on animals ranging in size from sardines to Blue whales.

matrilineal group consisting of a mature female and her offspring, and often their offspring. Such groups of resident animals form a series of progressively larger groups, up to about 60 individuals. Transients form much smaller groups of up to four, consisting only of a female and her immediate offspring. So far there have been no studies of social organization in Antarctic Orcas, but there is evidence that the different types exist there as well.

Females are physiologically capable of bearing a calf every three years, but they normally produce one only every eight years. The young measure 2 meters (7 ft) at birth, and calves remain in their birth pod with their mother for many years. Females generally live for 80 to 90 years and bear about five calves; males have a shorter lifespan of about 60 years.

FEMALE AT THE CENTER

This Orca can be identified as an adult female by the size and shape of its dorsal fin. An individual whale can be recognized by the unique shape of its saddle—the patch of white pigment behind the dorsal fin. Adult females are at the center of Orca society. The smallest stable groups consist of an adult female and her calves, some of them with calves of their own.

up to one third of its body length, and which is most fully developed in mature males. The blow of the Sperm whale has a forward angle of 45°, a feature which readily identifies it at sea. It has no teeth in the upper jaw and a relatively short, narrow lower jaw. The teeth in the lower jaw are the largest functional teeth known in the animal world, and may be up to 25 centimeters (10 inches) long. A particularly well-grown male Sperm whale may weigh three times as much as a mature female and can grow to twice the length. As well as being much smaller than the males, female Sperm whales have a less well developed spermaceti organ —the huge melon that gives the male such a distinctive square shape.

▲ **AVOIDING THE BENDS**
Before a feeding dive, Sperm whales lie at the surface blowing repeatedly—roughly one blow for every minute they will be submerged. Dives can last well over an hour. Although the whale plunges to extraordinary depths, it does not get the bends because it does not breathe oxygen under pressure. Its lungs empty soon after diving, and it relies instead on oxygen stored in blood and muscle.

SPERM WHALES

There are two families of sperm whales. The Sperm whale—the largest of all toothed whale species—belongs to the family Physeteridae; two other species, which are anatomically similar but much smaller, the Dwarf and Pygmy sperm whales, belong to the family Kogiidae. These two are as small as dolphins but several characteristics make it clear that they are more closely related to the Sperm whale than to any of the smaller toothed whales.

The most significant shared feature of the three species is the spermaceti organ (a highly specialized melon) which gives sperm whales their distinctive blunt profile. They also share highly unusual nasal passages. In all three whales the blowhole is located far to the left side of the head, but in the two smaller species it is at the back of the head, close to the normal position, whereas in the Sperm whale it is at the front of the melon. All three species have functional teeth in their small lower jaw, and all are very deep divers, feeding on squid that live at great depths.

Sperm whale

FAMILY: Physeteridae
PHYSETER MACROCEPHALUS
OTHER NAMES: Catchalot, Spermaceti whale
LENGTH: female 13 m/43 ft; male to 20 m/65 ft
WEIGHT: female up to 20 tonnes; male up to 50 tonnes
DISTRIBUTION AND HABITAT: Worldwide; observed in all the world's seas and oceans from the Antarctic through cold and temperate zones to tropical waters.
STATUS: unknown population, probably abundant
IUCN: Vulnerable   CITES: Appendix I
APPEARANCE: Sperm whales are very dark gray-brown with a wrinkled appearance over much of their body, and characteristic white scarring from deep-water squid.

One of the largest extant whales, the Sperm whale has a dorsal hump, rather than a dorsal fin. Its most recognizable feature is its huge, rectangular head, which may be

The largest brain on earth

Behind the spermaceti organ, the Sperm whale's huge skull contains a brain that can weigh more than 9 kilograms (20 lb). In comparison, the weight of a human's brain is only 1.3 kilograms (3 lb). Relative to its body size, the Sperm whale's brain has probably been as large as this for more than 30 million years, whereas humans reached their present brain capacity only 100,000 years ago. The whale's brain is very convoluted, with a highly developed cerebral cortex. This has been cited as evidence of high intelligence. However, intelligence is extremely difficult to compare across cultures, and is probably impossible across species. While many scientists believe that the size of a sperm whale's brain may be necessary for the animal to deal with the complex echolocation information that it must process, it is not possible to rule out intelligence of a different order than our own.

The skull of a Sperm whale is unusually asymmetrical for a mammal. The spermaceti organ in its forehead is filled with a huge quantity of waxy oil, and nasal passages run between air sacs at the front and back of this organ. Spermaceti oil was the most highly prized product of the nineteenth century whaling industry (except for ambergris, an aromatic waxy deposit that is also obtained from sperm whales). It was used as a fine machine oil, and is still unrivaled as a lubricant for missile inertial guidance systems.

Sperm whales use low-frequency clicks for echolocation or communication. Clicks may be arranged into organized patterns known as "codas," which are thought to mediate social interactions between animals that come together after a separation. They also produce "creaks"— very fast click trains rising in frequency—which may be used to track prey at close range. Whereas the sounds of Sperm whales appear relatively simple when compared with the songs of Humpback whales, their giant brains are probably capable of processing information of a richness and complexity that we cannot imagine.

Hunting

Sperm whales normally swim at 10 kilometers per hour (6 mph), but can reach speeds of up to 30 kilometers (20 mph). In the summer, mature males migrate slowly to polar waters where they live as solitary hunters, diving in search of squid to depths unmatched by any other mammal. They may ambush or pursue their prey through the depths before surfacing. Sperm whales eat mainly squid and octopus, but they will also opportunistically grab any deep-water species, including sharks and fish. They have been tracked to depths of 2,500 meters (8,000 ft), and there is evidence that they dive even deeper after prey. They usually patrol waters that are too deep for light to penetrate, and totally blind sperm whales have been captured in perfect health, and with food in their stomachs. Clearly these whales must have evolved a highly effective technique for detecting and catching their prey without the aid of vision, and echolocation is the obvious answer.

Family life

Female and young Sperm whales remain in nursery groups of some 20 to 40 whales, which are usually related through the mature females, and stay in temperate and tropical waters. These whales have very low reproductive rates, calving in summer or autumn and producing one offspring every four to six years. The young are 4 meters (13 ft) at birth. They are weaned into juvenile pods, but females may suckle their last offspring for up to fifteen years. As the whales mature, females rejoin their nursery groups, and males form bachelor herds that contain fewer numbers as the whales grow, until really large males tend to be very widely spaced, though they may still communicate with a group. During breeding, each nursery group is joined by up to five mature males, about one male for every ten females. Recent research suggests that males stay with a nursery group for only a few hours and then depart in search of another group where the females are ready to mate. Males compete for females, and only 25 percent are successful.

Beaked whales

The scientific name for beaked whales comes from the Greek word *xiphos*, meaning sword. Beaked whales are small to medium species. They have a single pair of throat grooves, unlike most toothed whales, which do not have any. They are a very successful group of whales, and there are around twenty species, most of which are less than 6 meters (20 ft) long.

Beaked whales are generally deep-water animals, and are very reclusive. Populations could be quite large, especially for Arnoux's beaked whale and the Southern bottlenose whale, which live in Antarctica in summer; however, they are seldom seen, and scientists know very little about their behavior. On occasions, they have been spotted in groups of fifty or more, which suggests that they are gregarious. Beaked whales feed on squid,

and while males have one or two tusk-like lower pairs of teeth, females have no real teeth. Males seem to use these teeth for fighting amongst themselves.

Arnoux's beaked whale

BERARDIUS ARNOUXII

OTHER NAME: New Zealand beaked whale

LENGTH: 10 m/33 ft

WEIGHT: 10 tonnes

DISTRIBUTION AND HABITAT: Circumpolar throughout the southern oceans.

STATUS: unknown population, probably common in Antarctica

IUCN: Lower Risk—conservation dependent

CITES: Appendix I

APPEARANCE: These whales are blue-gray or brownish, with slightly darker flippers, flukes, and back. Older males tend to be paler and can be heavily scarred. This is one of the few beaked species with functional teeth.

Southern bottlenose whale

HYPEROODON PLANIFRONS

OTHER NAMES: Flower's/Flat-headed/Antarctic bottlenose whale, Pacific beaked whale

LENGTH: 7.5 m/25 ft

WEIGHT: 8 tonnes

DISTRIBUTION AND HABITAT: Throughout the southern hemisphere, particularly in Antarctic waters, but occasionally reaching into the tropics.

STATUS: unknown population, thought to be common in Antarctica

IUCN: Lower Risk—conservation dependent

CITES: Appendix I

APPEARANCE: Southern bottlenose whales are an unusual metallic gray on top and paler below, and they may be heavily scarred. Females are larger than males. LW

▼ MYSTERY WHALE
Little is known of the Southern bottlenose whale. Although it is believed to be reasonably common in the Southern Ocean, it is rarely seen, and is difficult to identify unless observed at close range. It dives deep to feed on squid, but its surface behavior is inconspicuous. These enigmatic animals are thought to migrate to warm waters during the winter months.

Birds
Seabirds

Of the nearly 10,000 bird species on earth, only about three percent are considered seabirds, meaning that they obtain virtually all their sustenance from the sea. Some 70 percent of the earth's surface is covered in water, and this is the world of seabirds.

Many seabirds and a few land birds make their home in Antarctica or on the many sub-Antarctic islands scattered around the Southern Ocean. To stand on the deck of a ship in the Drake Passage, the waterway that separates the southern tip of South America from the Antarctic Pensinsula, is to marvel at the birds that wheel and dive above the waves. Humans cannot imagine a life away from *terra firma*—so how do these birds survive on an apparently featureless ocean?

At home at sea

Unlike marine mammals, no seabirds have evolved the ability to reproduce without coming ashore somewhere to find a mate, build a nest, and lay eggs. Most seabird species breed only once each year (some only every other year), spending four to five months in the process. Tied to land for less than half of each year, seabirds are truly at home when at sea. Most seabird species are relatively long-lived and take a long time to reach sexual maturity, so that once a young bird fledges and strikes out on its own, it may be years before it alights on land again. There is no fresh water at sea, and food, while abundant, is often found in widely separated patches or hidden deep beneath the surface. Finding ways to meet these

◁ THE NEXT GENERATION
A White-capped albatross (*Diomedea cauta*) gently preens its small chick. Once the egg hatches, the parents tend the chick for about three weeks. During that time the male and female take turns going to sea for one to three days in search of food while the other feeds, broods, and guards the chick.

▲ BREEDING COLORS
In the midst of the breeding season, the Imperial shag (*Phalacrocorax atriceps*) sports a bright blue eye-ring and yellow caruncles at the base of its bill. It builds a substantial nest from seaweed and pieces of grass cemented together with guano.

◥ DISTINGUISHING TRAITS
A bloodstained Southern giant petrel (*M. giganteus*; bottom) and a Northern giant petrel (*Macronectes halli*; top) scavenge an elephant seal carcass. The Northern giant petrel has a characteristic red tip on its bill; only Southern giant petrels have an all-white form.

challenges has given rise to the diversity of seabirds in the world today.

Several groups of birds have adapted to life at sea, and there is no single story of seabird evolution. Of the flying birds around the world, 113 species are Procellariiformes, or tube-nosed birds, expert fliers that roam huge expanses of the oceans to find food; they are an old group with great diversity that evolved in the southern hemisphere. Descended from ancestral Procellariiformes, another 50 species of seabirds are the Pelecaniformes, the pelicans, gannets and boobies, tropicbirds, cormorants and shags, and frigatebirds; these tend to live in temperate or tropical habitats, and stay closer to shore than the tube-nosed seabirds. Another 107 species come from a larger group of waterbirds, the Charadriiformes; these are the skuas, gulls, terns and noddies, skimmers, and auks. Primarily northern hemisphere birds, they have spread throughout the world's seas. Several other scattered species can also claim the title seabird: a few species of ducks and geese of the order Anseriformes, and grebes (Podicepediformes) and loons (Gaviiformes).

Seabirds wander more than land birds, and transequatorial crossings have occasionally led to the rise of closely related species pairs, with one species in the north and the other in the south—for example, Arctic/Antarctic terns, Lesser black-backed/Kelp gulls, Great/Antarctic skuas, and Northern/Southern fulmars. Some of these pairs are literally sister species while others are the result of convergent evolution.

Eliminating salt

Seabirds face a problem not usually encountered by land birds: when they spend months at sea, there is no fresh water for them to drink. It was known for many years that bird kidneys were less efficient than human kidneys, and no one understood how birds dealt with excess salt. Anatomists knew that all seabirds have two small glands lying in a small groove above their eyes, but it was not until the late 1950s that animal physiologist Knut Schmidt-Nielsen found the explanation: the function of these glands is to get rid of excess salt. The salt glands are ten times more efficient at removing salt than bird kidneys. Salt is picked up by the circulating blood, and as it passes through the salt glands it is excreted as a highly saline solution that drains into the nasal cavities. In most birds this solution drips from the nostrils to the end of the bill, and the characteristic head-shaking of most seabirds is to get rid of these drips of salt. Many kinds of land birds possess rudimentary salt glands, but they are tiny compared with those of seabirds.

Albatrosses

Albatrosses have an almost mythic significance for sailors: early venturers into the Southern Ocean believed that to shoot an albatross would bring bad luck, and biologist Robert Cushman Murphy said: "I now belong to the higher cult of mortals, for I have seen the albatross." Despite their remote locations, commercial interests have imperiled albatrosses for many years. In the late nineteenth and early twentieth centuries, many species were brought to the verge of extinction by hunters for the feather trade. Today thousands of albatrosses are hooked and drowned when they try to seize baits from the hooks of longline fishing operations.

The distinctive feature of albatrosses is their size: the Wandering albatross, with a wingspan of up to 3.5 meters (11½ ft), is the largest flying bird in the world. Albatrosses belong to two genera in the family Diomedeidae: the 12 *Diomedea* species, and the two smaller and darker *Phoebetria* species. The Wandering and Royal albatrosses are popularly known as the great albatrosses, and the smaller *Diomedea* species as mollymawks.

Albatrosses have long wings relative to their body size, and their vertebrae and upper wing bones are hollow yet strong, reinforced with struts to reduce their body weight and make gliding easier. Even with these adaptations their long wings are awkward, and they are not very maneuverable. They soar majestically over the waves, seldom flapping their wings. Albatrosses are the most oceanic of the seabirds and cover vast distances. Even in the breeding season, when they must return regularly to their nest to feed chicks, they may travel 8,000 kilometers (5,000 miles) in a week in search of food.

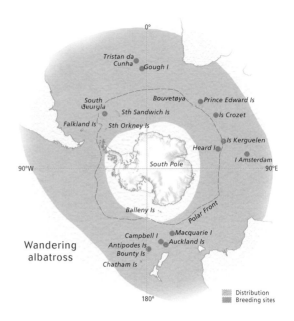

Wandering albatross

Distribution
Breeding sites

Wandering albatross

DIOMEDEA EXULANS

LENGTH: 107–135 cm/42–53 in
WINGSPAN: 250–350 cm/98–138 in
STATUS: about 21,000 breeding pairs; 100,000 birds
IUCN: Vulnerable   CITES: not listed

The Wandering albatross is the largest of the albatrosses. As a juvenile its feathers are brown but as it matures, it becomes progressively whiter. First the body whitens and then the upper wings, starting with white spots on the mid-wing close to the body and slowly whitening along the wings to cover the entire wing except the tips. The tails retain some black.

The breeding cycle occupies just over a year, and the birds spend the year following breeding at sea: so at best they can raise one chick every two years. They breed in loose aggregations on gentle hills with good exposure to the wind. Eggs are incubated for 78 to 79 days before the fluffy white chick hatches. Chicks stay in the nest and are fed by both parents for an average of 246 days before they are ready to go out on their own. Towards the end of the nestling period, the older fledglings leave the nest and walk around, exercising their wings.

Although traditionally considered a single species containing five subspecies, recent genetic studies argue that the Wandering albatross is actually five distinct species, each with unique characteristics in its plumage.

Wandering albatrosses feed primarily on squid and fish, which they seize once they land on the water surface. They eat more carrion than other albatrosses, and are habitual ship-followers.

Royal albatross

DIOMEDEA EPOMOPHORA

LENGTH: 107–122 cm/42–48 in
WINGSPAN: 305–350 cm/120–138 in
STATUS: about 10,000–20,000 breeding pairs
IUCN: Vulnerable   CITES: not listed

Closely related to the Wandering albatross, the Royal albatross is very similar in appearance. Immature Royal albatrosses, however, are mostly white, and only young juveniles have black on their tails. Royal albatrosses also have distinctive black lines along the cutting edges of their upper bill, although these can only be seen at close range. The adults of the southern subspecies retain the dark tops of the wings.

The breeding behavior of the Royal albatross is exactly like that of the Wandering albatross, although the precise timing may differ according to the island where they breed and the conditions that they find there. Like Wandering albatrosses, Royal albatrosses feed extensively on squid, which they take by surface-seizing. Unlike Wandering albatrosses, they are not habitual ship-followers, but they readily join other petrels to feed on offal discarded from fishing boats.

RESTING AT SEA
RESTING AT SEA
A Yellow-nosed albatross swims calmly on the sea. The usual image of albatrosses is soaring on the wind, but they regularly alight on the water to feed or rest. They have no trouble taking off as long as there is a wind blowing.

Yellow-nosed albatross

DIOMEDEA CHLORORHYNCHOS

OTHER NAME: Yellow-nosed mollymawk
LENGTH: 71–81 cm/28–32 in
WINGSPAN: 200–256 cm/79–101 in
STATUS: 80,000–100,000 breeding pairs
IUCN: Lower Risk–near threatened
CITES: not listed

The Yellow-nosed albatross, with its white body and dark wings, looks like the other mollymawks, but its gray head is lighter than that of the Gray-headed albatross, and it has a whitish cap. Its underwings have more white than those of the Black-browed and Gray-headed albatrosses but not as much as those of the White-capped albatross. It is more slimly built than other mollymawk species and its long, thin bill is black, with pronounced yellow lines above and below.

Yellow-nosed albatrosses start their annual breeding cycle in August and September. They nest singly or in loose groups on fairly flat ground, in nests of mud and grass. Parents incubate the single egg for 71 to 78 days, and then feed the chick for about 130 days, when it fledges. After the breeding season they disperse around the South Atlantic and Indian oceans between latitudes 15 and 50°S.

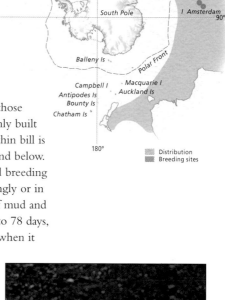

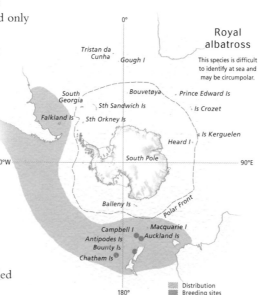

EGG LAYING
A female Royal albatross lays its egg in mid-November. The egg takes about 40 days to form in the female, but laying typically lasts just three minutes.

Black-browed albatross

DIOMEDEA MELANOPHRIS

OTHER NAME: Black-browed mollymawk

LENGTH: 83–93 cm/32–36 in

WINGSPAN: 240 cm/94 in

STATUS: 550,000–600,000 breeding pairs;
more than 2 million birds

IUCN: Lower Risk–near threatened

CITES: not listed

Black-browed albatross

Distribution
Breeding sites

The Black-browed albatross is probably the most common of the albatrosses and one of the most recognizable. Its body is white with a black saddle between its wings, its upper wings are black, and its underwings are white with a broad black margin. It has a yellowish bill, and a white head with clearly marked black eyebrows. From a distance, and at rest, it can be mistaken for the Kelp gull, but its soaring flight, in elegant contrast to the flapping motion of the Kelp gull, identifies it as an albatross. It feeds on crustaceans—mainly krill—as well as on squid and fish, but it also devours carrion, joining other petrels at fishing boats to scavenge the offal.

Black-browed albatrosses breed annually, the females laying a single egg in a substantial nest of mud and grass. Breeding colonies can consist of as many as 100,000 pairs of birds, and are often shared with shags and penguins.

Little is known of their migratory patterns after breeding, but there is a general movement northward in the southern autumn. Stragglers are regularly seen in the North Atlantic, and there have been over 40 recorded sightings in Britain.

White-capped albatross

DIOMEDEA CAUTA

OTHER NAMES: Shy albatross, Mollymawk

LENGTH: 90–99 cm/35–39 in

WINGSPAN: 220–256 cm/87–101 in

STATUS: 800,000–1,000,000 birds

IUCN: Lower Risk–near threatened

CITES: not listed

White-capped albatross

Distribution
Breeding sites

The White-capped albatross is the heaviest and most thickset of the mollymawks. The undersides of its wings are almost entirely white, with a thin border of black, and it has characteristic black "thumbprints" on the underwing near the body. Its head is a very pale gray, and its dark eyes and eyebrows make it look stern. Its bill is light gray with a yellowish tip.

White-capped albatrosses normally breed annually, starting in September, and their

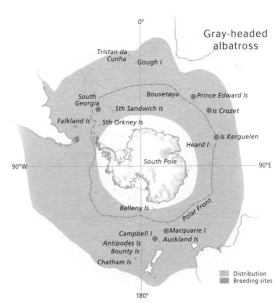

▲ SLOW AND STEADY

Gray-headed albatrosses (*D. chrysostoma*) are biennial breeders. Their chicks grow slower and are fed longer than other mollymawks, but they do have consistently higher fledging success.

Gray-headed albatross

Distribution
Breeding sites

breeding cycle is the same as those of other mollymawks. They eat mostly squid, as well as fish, crustaceans, and offal from fishing boats. Despite one of their common names—Shy albatross—they often feed with other seabirds, and they have been seen following whales. They compete successfully with the birds that crowd around fishing boats.

Gray-headed albatross

DIOMEDEA CHRYSOSTOMA

OTHER NAME: Gray-headed mollymawk

LENGTH: 81–84 cm/32–33 in

WINGSPAN: 180–220 cm/70–87 in

STATUS approximately 500,000 birds

IUCN: Vulnerable   CITES: not listed

Except for its gray head, which gives it a hooded appearance, this species is very much like the Black-browed albatross. Its bill is black with yellow lines along the top and bottom. Its flight patterns, food, and lifestyle are similar to those of the Black-browed albatross, and the two species often share colonies, although the Gray-headed albatross prefers steeper slopes. The structure and timing of the reproductive cycle are also similar, except that Gray-headed albatrosses feed their chicks for three or four weeks longer.

Gray-headed albatrosses prefer colder water than most other albatrosses, and they stay well away from land except at their breeding sites. They eat mostly squid, and they follow fishing vessels less than other albatrosses.

Sooty albatross

PHOEBETRIA FUSCA

OTHER NAME: Dark-mantled sooty albatross

LENGTH: 84–89 cm/33–35 in

WINGSPAN: 203 cm/80 in

STATUS: 80,000–100,000 birds

IUCN: Vulnerable   CITES: not listed

Light-mantled sooty albatross

PHOEBETRIA PALPEBRATA

LENGTH: 78–79 cm/30–31 in

WINGSPAN: 183–218 cm/72–86 in

STATUS: about 150,000 birds

IUCN: Lower Risk–near threatened CITES: not listed

The two albatrosses that constitute the genus *Phoebetria* can be distinguished from other albatrosses by their narrow, pointed wings and long, wedge-shaped tails. Returning to shore only during their breeding period, they nest in loose colonies among vegetation on cliffs or steep slopes. The breeding areas of the two species overlap in some areas, and where this happens they lay their eggs at different times. They feed mainly on squid, but they also take fish, crustaceans, and carrion, usually penguin carcasses.

Sooty albatrosses are uniformly dark, with a partial white eye ring and a yellow line along the cutting edge of their lower bill. Little is known of their feeding habits, but they probably surface-feed at night. Sooty albatrosses on Nightingale Island in the Tristan da Cunhas are under threat from humans; each year as many as 3,000 chicks are taken from a major colony there, and the population cannot withstand such heavy exploitation.

Identical to the Sooty albatross except for an ash-gray collar and mantle and a blue line on the lower mandible, Light-mantled sooty albatrosses prefer colder water than do the Sooty albatrosses. They also like to breed further inland.

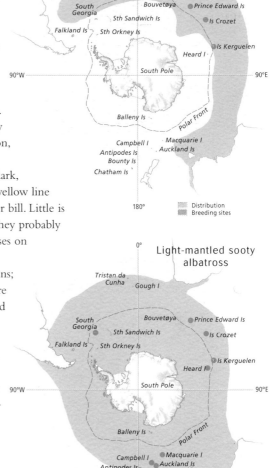

Sooty alba-
tross

Distribution
Breeding sites

Light-mantled sooty
albatross

Distribution
Breeding sites

▼ ELABORATE RITUALS

Light-mantled sooty albatrosses court with alternating aerial and ground-based displays. At the nest, courtship proceeds with a series of behaviors such as mutual preening, open-mouthed thrusting, tail fanning, and bowing.

▲ LARGE BABIES

Light-mantled sooty chicks are guarded and fed by their parents for the first 19–21 days. They grow up to 1.4 times heavier than their parents, but by the time they make their first flight they dwindle to 91 percent of typical adult weight.

Fulmars

This family consists of a diverse group of petrels—small to medium-sized seabirds, dark in color, with long wings that enable them to use the wind to fly great distances. The nostrils on the top of their bill join into a single tube with a visible septum separating them. All petrels lay only one egg at each breeding attempt, but incubation times are quite variable. They often leave their egg unattended while they forage. Because of these breaks in incubation, petrel embryos withstand cold better than those of most other birds. The period from hatching to fledging also varies due to changes in local feeding conditions and the parents' ability to raise their chick.

Family origins

The first group of petrel-like birds diverged from the divers, or loons (Order Gaviiformes) about 50 million years ago, and soon after the Diomedeidae (albatrosses) diverged from the petrels. The first fossil evidence of the Procellariidae dates from 40–50 million years ago and is considered to be from a shearwater. They probably originated in the southern hemisphere, later spreading northward. Petrels are now found in all of the world's oceans, but the southern hemisphere still supports many more petrel species than the northern hemisphere.

This group presents many taxonomic puzzles that are exacerbated by lack of information about their breeding behavior and ecology. However, the family can be split into four main groups based on anatomical and ecological differences such as their flight patterns, feeding habits, and bill shapes; they are the fulmarine petrels, the prions, the gadfly petrels, and the shearwaters.

Fulmarine petrels

Ranging from the Arctic to the Antarctic, the fulmarine petrels are the most diverse of this group. All fulmarine petrels have relatively long, broad wings, but they vary considerably in size and plumage. There are two sets of sibling species—the Northern and Southern giant petrels, and the Northern and Southern fulmars—and three further species: the Snow petrel, the Antarctic petrel, and the Cape petrel. The giant petrels fly with slow, powerful wingbeats, while the smaller species have faster wingbeats; all are good gliders. Fulmarine petrels feed on squid and crustaceans, mainly surface krill. The giant petrels, fulmars, and Cape petrels are ship-followers.

Prions

Prions are small petrels of the genus *Pachyptila*. Prion species include the Narrow-billed, Antarctic, Broad-billed, Fairy and Fulmar prion, plus the Blue petrel—the only member of the genus *Halobaena*. (The Blue petrel has been placed in each of the four Procellariid groups, but its coloration and other structural features link it with the prions.) Prions are similar in appearance and habits, sharing a buoyant, erratic flight. All are a light blue-gray with white underparts, dark primary feathers, and upper wing coverts forming an M-shape on their upper wings. All have black tips on their wedge-shaped tails, except for the Blue petrel, which has a white tip. Some breeding sites are threatened by predation by introduced rats, cats, and pigs, and commercial krill harvesting may limit food supply.

Gadfly petrels

Traditionally this group comprises 23 to 34 *Pterodroma* species and two *Bulweria* species, but relationships within the group are not understood. Only a few *Pterodroma* species are cold-climate birds, including the Great-winged, White-headed, Kerguelen, Soft-plumaged and Mottled petrels. Several of the *Pterodroma* species are known only from a few specimens taken long ago and only recently rediscovered. Gadfly petrels have long, wedge-shaped tails and relatively short wings. In flight they flap and glide, rising in high,

swooping arcs, and they tend to bend their wrists while flying. They rarely land on water, but pluck squid, crustaceans, and fish from the surface; they are not ship-followers, and they do not dive for food. Most nest in small colonies or scattered pairs, but because of their isolated breeding areas and their secretiveness, the breeding sites of several species are unknown. Many colonies of gadfly petrels are threatened by predation and harassment by introduced species.

Shearwaters

Shearwaters are a very old and complex group consisting of four species in the genus *Procellaria*, some 15 to 20 in *Puffinus*, and two in *Calonectris*—the latter two species rarely venturing into Antarctic waters. The range of many shearwaters is only partially in Antarctic or sub-Antarctic waters, and several migrate to northern hemisphere waters. Southerly-ranging *Puffinus* species include the Sooty, Great and Little shearwaters. They are medium-sized to large petrels with long bills, long, narrow wings, and short, rounded tails. The smaller shearwaters alternate bursts of rapid wingbeats with stiff-winged banking glides near the ocean surface, dipping into the troughs and rising over the crests of the waves. The flight of the larger species is more relaxed and deliberate. Their plumage is dark above, and brown or white below. They feed on fish, crustaceans, and squid on the surface or from shallow dives. Only the four *Procellaria* species regularly venture into Antarctic waters, with the White-chinned and Gray petrels more common and widely distributed than the Westland and Parkinson's petrels, and none of the species has a wholly Antarctic distribution.

> **ARMED AGAINST ATTACK**
Southern fulmars nest in the open on cliff ledges and steep slopes. They can protect themselves and their chicks from attacks, often by skuas, by spitting or spraying sticky stomach oil onto attackers. They will even spit on people who make them nervous by approaching too close.

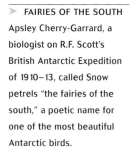

> **FAIRIES OF THE SOUTH**
Apsley Cherry-Garrard, a biologist on R.F. Scott's British Antarctic Expedition of 1910–13, called Snow petrels "the fairies of the south," a poetic name for one of the most beautiful Antarctic birds.

▽ **OCEAN SCAVENGERS**
A Northern giant petrel feeds on a dead penguin in shallow water. The giant petrels, especially the males, are the top scavengers throughout the Southern Ocean. They dominate all other petrels and skuas at a carcass.

Southern giant petrel

MACRONECTES GIGANTEUS

OTHER NAME: Southern giant fulmar

LENGTH: 86–99 cm/34–39 in

WINGSPAN: 185–205/73–81 in

STATUS: approximately 36,000 breeding pairs

IUCN: Vulnerable   CITES: not listed

The Southern giant petrel, with its close relative, the Northern giant petrel, is the largest of the Procellariidae. These species are sometimes mistaken for small albatrosses, but they do not glide as much as albatrosses, their bodies are heavier and more hunched, and they have thicker bills because of the size of their tubular nostrils on the top of the bill.

Southern giant petrels have heavy, yellow-green bills with pale green tips. Most adults are brownish gray, with dirty white heads, necks, and upper breasts; juveniles are sooty black with a few white flecks on the head. Their circumpolar range extends south to the edge of the Antarctic ice and north into the subtropics. They breed on exposed hilltops and plains on sub-Antarctic islands throughout the region in small, loose colonies, building nests of small stones or grass. Breeding starts in October, with both parents taking turns to incubate a single egg for 55 to 66 days.

Giant petrels are scavengers, feeding on carrion, especially the carcasses of seals,

penguins, and petrels. They can be aggressive, and will drown or batter to death such birds as Cape petrels or immature albatrosses. They also surface-feed on squid, krill, and fish. They are avid ship-followers in quest of refuse or offal, and many are caught on the hooks of longline fishing boats.

Antarctic petrel

THALASSOICA ANTARCTICA

LENGTH: 40–46 cm/16–18 in

WINGSPAN: 101–104 cm/40–41 in

STATUS: many millions of birds

IUCN: not listed   CITES: not listed

This is one of the most numerous of the Antarctic birds. Like the Cape petrel, for which it is often mistaken from below, it has a white underbody and wings and a black face and throat. However, its upper side is quite distinctive. Its head, body, and the front half of its wings are dark brown to dark gray, while the back half of its wings are white, as is its tail except for a black tip.

Antarctic petrels are hard to observe because they stay mostly within the boundaries of the pack ice, although some do fly as far north as the Polar Front in winter. They are truly Antarctic birds, feeding in open water near pack ice and favoring areas with icebergs. They are sometimes found in association with other seabirds and whales, and will sometimes roost in large numbers on icebergs. They nest on cliff ledges and begin breeding in November. About 35 nesting colonies have been identified, some as far as 250 kilometers (155 miles) inland, but the known colonies cannot

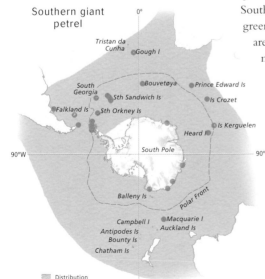

Southern giant petrel

Tristan da Cunha
Gough I
Bouvetøya
Prince Edward Is
South Georgia
Sth Sandwich Is
Is Crozet
Falkland Is
Sth Orkney Is
Is Kerguelen
Heard I
90°W
South Pole
90°E
Balleny Is
Polar Front
Campbell I
Macquarie I
Antipodes Is
Auckland Is
Bounty Is
Chatham Is
0°
180°

Distribution
Breeding sites

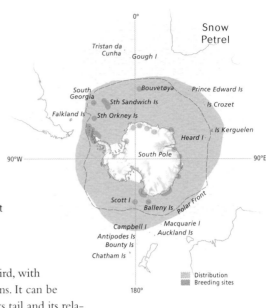

Snow Petrel

◄ INLAND NESTERS
Antarctic petrels usually feed in open water around the pack ice, but they nest on cliff ledges, sometimes more than 200 kilometers (120 miles) from the coast.

Antarctic prion

PACHYPTILA DESOLATA

LENGTH: 25–27 cm/10–11 in
WINGSPAN: 58–66 cm/23–26 in
STATUS: many millions of birds; abundant and widespread
IUCN: not listed   CITES: not listed

The Antarctic prion is a fairly large bird, with the widest distribution of all the prions. It can be identified by the broad black tip on its tail and its relatively extensive dark blue-gray collar. It feeds mainly on krill and other small crustaceans, although it will also take some fish. Its bill is adapted for filter feeding, and it has developed a specialized behavior, called hydroplaning, to take advantage of this feature: it scurries along the surface of the water into the wind, with its body resting on the water but partially supported by the wind flowing under its outstretched wings. Holding its mouth open at the surface, or with its head entirely submerged, it filters its prey from the water as it swims, sometimes swinging its head from side to side.

Antarctic prions range from latitude 40°S to the ice pack, and are circumpolar except for a gap in the southern Pacific. They breed on islands from New Zealand westward to South America, nesting in burrows in very dense colonies. They usually begin breeding in November—later if there is too much snow on the ground. After their chick fledges, at 45 to 55 days, all the birds leave the colonies and disperse northward over a broad area.

Despite the large numbers of Antarctic prions, they have problems in some of their breeding areas from introduced rats, cats, and pigs. Large-scale harvesting of krill may cause concern in the future.

account for their estimated numbers. They feed mainly on krill, and to a lesser extent on fish and squid, which they take by surface-seizing, although they also plunge-dive and dip for food.

Snow petrel

PAGODROMA NIVEA

LENGTH: 30–40 cm/12–16 in
WINGSPAN: 75–95 cm/29½–37 in
STATUS: possibly several million birds
IUCN: not listed   CITES: not listed

Snow petrels have pure white plumage, dark eyes, and a dark bill. They forage in areas with pack ice. They dip-feed on the wing, taking mainly krill with some fish, squid, and carrion. They also land on the surface and feed by surface-seizing or shallow diving. They are circumpolar, nesting extensively on the Southern Continent and on some of the islands in high latitudes. Some remain around the Continent all year, but the main influx of breeders arrives from mid-September to early November.

Snow petrels, along with South polar skuas, breed further south than any other birds in the world, in some of the harshest conditions on earth. They nest in small to large colonies in crevices in cliffs or among boulders on scree slopes, mostly near the coast, but sometimes as far as 345 kilometers (215 miles) inland, at elevations as high as 2,400 meters (7,870 ft). In the bitter weather they must dig the snow out of the crevices where they nest. Incubation averages 45 days, and the young are brooded for eight days after hatching; the parents then go to sea to bring food back to their chick. Chicks leave the colony at 42 to 50 days old.

Antarctic prion

▼ DISTINCTIVE PLUMAGE
Unmistakably a prion, with its bluish back and the bold "M" pattern formed by black primaries and diagonal black stripes across the top of the wings, this Antarctic prion also has a strong black eye line and an extensive dark collar.

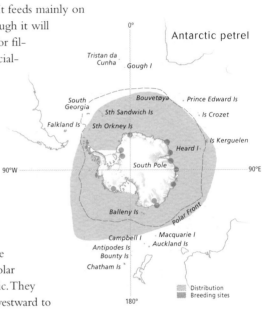

Diving-petrels

There is sparse fossil evidence of the origins of diving-petrels, but they probably evolved in isolation somewhere around southern South America or Australia, and then spread around the Southern Ocean. Whereas most tube-nosed seabirds are supremely adapted to an aerial life on the open ocean, diving-petrels have evolved in a different direction; they have adapted to flying under water, and could be considered flying penguins. To adapt to swimming, diving-petrels developed heavy, stocky bodies with short, rounded wings. Like those of most aquatic birds, their feathers are dark above and light below, with very little pattern. Their short, black bills have a small elastic pouch like that of a pelican for food storage—hence their family name, Pelecanoididae. To counter buoyancy in the water, their bones are denser than those of their more aerial cousins.

These adaptations act against them when they fly, so they must work very hard when in the air. Most petrels glide on wind currents, flapping their wings intermittently, but diving-petrels are a constant whir of wings, and their bones are so dense that they would drop like stones if they stopped flapping. Their wings beat so quickly that they appear as a blur alongside a small bullet-shaped body. They normally fly short distances, and very close to the surface. They do not skip and dance over the waves like a shearwater, but fly directly into waves and pop out on the other side.

Diving-petrels are strictly marine, but because they need so much energy to fly, they are most often found inshore rather than far out to sea. Most tend to stay close to their breeding areas. At sea they are usually seen singly or in small groups, roosting on the water when not actively feeding.

Diving-petrels nest in burrows that they dig on sloping ground, sometimes in colonies of thousands of pairs. Normally they come ashore under cover of darkness to avoid the gulls and skuas that prey on them. They spend the night inside their burrows, calling during the night. During the breeding season, there are always many birds flying overhead. Once paired, the females spend some time feeding intensively at sea, then lay a single egg. Both sexes take turns to incubate the egg for seven or eight weeks in daily shifts—the shortest incubation shift of any of the tube-nosed seabirds—because their feeding areas are so close to their breeding grounds. Parents feed the chick for about eight weeks before it goes to sea on its own. Chicks that survive their very difficult first year join the flocks of immature birds wheeling over the breeding colony in the following year.

▼ BUSY BIRDS
Common diving-petrels seemingly stop only when they arrive at their nest burrows at night. At sea they are a constant whir of wings as they dive through the waves and pursue their prey. They will rest on the water, but dive quickly out of sight when a boat approaches.

South Georgia diving-petrel

PELECANOIDES GEORGICUS
LENGTH: 18–21 cm/7–8 in
WINGSPAN: 30–33 cm/12–13 in
STATUS: more than 4 million breeding pairs
IUCN: not listed   CITES: not listed

The South Georgia diving-petrel has a wide-based bill that tapers in a continuous curve to a point. It breeds on most sub-Antarctic islands; on Iles Kerguelen, where its range overlaps with that of the Common diving-petrel, the South Georgia species forages further from the island. It feeds mainly on planktonic crustaceans, especially krill and amphipods, as well as some small fish and squid. There are more than 2 million pairs on South Georgia.

Common diving-petrel

PELECANOIDES URINATRIX
LENGTH: 20–25 cm/8–10 in
WINGSPAN: 33–38 cm/13–15 in
STATUS: more than 4 million breeding pairs
IUCN: not listed   CITES: not listed

The Common diving-petrel's bill looks long and narrow because the sides are parallel along the length of the bill until near the tip. The species is found throughout the sub-Antarctic region. The Kerguelen diving-petrel (*P. u. exsul*), one of six subspecies of the Common diving-petrel, ranges from South Georgia east to the Antipodes Islands. It is most abundant around Iles Kerguelen, where there may be as many as one million breeding pairs.

Storm-petrels

Storm-petrels probably diverged quite early from the other petrels. All storm-petrels probably originated in the southern hemisphere and colonized the north.

About 18–20 centimeters (7–8 in) long, storm-petrels have shorter and broader wings than other petrels. They dart and weave at wave level in agile and restless flight. In a strong wind they may glide slowly, with wings raised in a V-formation and legs dangling. They have highly developed salt glands, long, tubular nostrils, and a keen sense of smell for detecting food. Most are black to dark brown on their upper parts, with some gray or white on the face, rump, tail, flanks, or belly. Their flight feathers are mainly dark because of a high concentration of the pigment melanin, which toughens their feathers to withstand salt and wind.

Although small, storm-petrels are usually found far out to sea except during the breeding season, and they are relatively long-lived, with life spans of six to twenty years. They inhabit all the world's seas except in the highest northern latitudes. They nest in remote, predator-free areas in cliff crevices, boulder fields, or scree slopes, usually coming to their colonies only at night. Colonies range in size from a few dozen nests to tens of thousands scattered over large slopes.

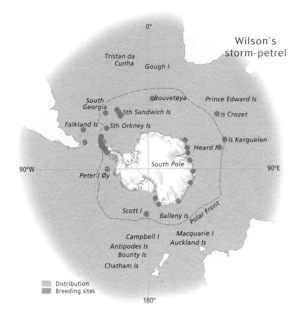

Wilson's storm-petrel

breeding season it flies to the North Atlantic, to the Gulf of Alaska in the Pacific Ocean, and throughout the Indian Ocean—amazing journeys for such small birds.

Other storm-petrels

Three other storm-petrels breed in Antarctic and sub-Antarctic waters. The Black-bellied storm-petrel (*Fregetta tropica*) is widespread in the Southern Ocean. It has the typical dark back and white rump, but the white extends to most of its belly and breast, with a thick black line bisecting the belly. It breeds on islands and isolated off-shore stacks to avoid predation by introduced cats, rats, and mice. In the non-breeding season it travels through-out the southern hemisphere, and may even reach the northern parts of the Indian Ocean.

The Gray-backed storm-petrel (*Garrodia nereis*) has a dark back, a white belly, and a gray band on its rump and tail. It occurs throughout the Southern Ocean, but is more common around New Zealand and the nearby sub-Antarctic islands in the Indian Ocean and western South Pacific. It stays closer inshore over the continental shelf than other storm-petrels, probably visiting deeper waters only when it disperses.

The White-faced storm-petrel (*Pelagodroma marina*) takes its name from the prominent white pattern on its face. Its back is gray-brown, and its underside mainly white. Six sub-species are scattered on remote island groups throughout the Atlantic, southern Pacific, and Indian oceans. Although they are very common— with about a million in New Zealand and its outlying islands alone—they are vulnerable to destruction of their breed-ing areas by livestock, and to exploitation by fishers. In 1970, over 200,000 were found dead on Chatham Island, their legs entangled in the filaments of *Distomium filiferum* (large flatworms).

▲ **BLACK STRIPE**
In seeming contradiction to their name, Black-bellied storm-petrels appear to have a white belly when seen on the wing. However, though it is hard to see in flight, they do have a thick black line down the center of their belly.

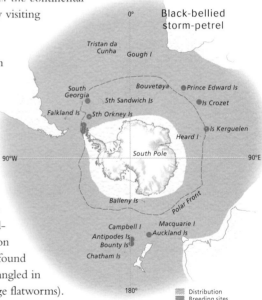

Storm-petrels are monogamous unless they repeated-ly fail to reproduce. Both sexes incubate the egg in stints of two to eight days, although they may leave the egg unattended for long periods of time. The chick fledges after seven to eleven weeks. Chicks that survive the first couple of years join with flocks of non-breeders that visit the breeding colony for several years before becom-ing mature enough to breed themselves.

Wilson's storm-petrel

OCEANITES OCEANICUS
LENGTH: 15–19 cm/6–7½ in
WINGSPAN: 38–42 cm/15–16½ in
STATUS: many millions of birds
IUCN: not listed   CITES: not listed

Wilson's storm-petrel is all dark with a simple white band across its rump, and sometimes lighter bars above and below the wings. In flight, its legs project beyond its tail. It prefers cold waters above continental shelves, feeding mostly on crustaceans such as krill, on floating carcasses, and on fish offal and rubbish from ships. It is a transequatorial migrant, with one of the longest record-ed migrations. It breeds only on sub-Antarctic islands and around the Antarctic Continent, but in the non-

RELYING ON BIGGER BIRDS
Gray-backed storm-petrels often attend the feeding sites of other birds—especially if there is a carcass floating in the water. As they are quite small, they cannot pull the carcass apart for themselves, but there are plenty of scraps to scavenge.

Sheathbills

Sheathbills are medium-sized, pigeon-like shorebirds that belong to the same order as skuas, gulls, and terns. They probably evolved around the tip of South America from a plover-like ancestor. Sheathbills are the only bird family found entirely within the Antarctic and sub-Antarctic regions. Today, there are just two species in the family, both of the genus *Chionis*. They are different in appearance but very similar in habits.

Both species inhabit coastal areas and rely on the sea for some of their foraging, feeding on algae, limpets, and other small invertebrates in the intertidal zone. They are the ultimate scavengers of Antarctica, eating virtually any form of organic matter. They forage among penguin and cormorant colonies for unattended eggs or small chicks, dropped food, the expelled stomach linings of penguins, and the feces of other birds and seals. They congregate around concentrations of seals in the early spring, feeding on placentas and the inevitable seal carcasses, picking at scabs and wounds on adult seals, and even feeding on the nasal mucus of elephant seals.

They are expert at stealing food from penguins, cormorants, and albatrosses by leaping against adults who are feeding their chicks and quickly scooping up spilled food. They are very agile, easily dodging penguins and seals that lunge for them as they forage. Their fearlessness also brings them into close contact with humans at settlements and scientific bases, where they scavenge absolutely anything they can find. Unfortunately, they sometimes eat dangerous items like grease from machinery or small batteries.

Sheathbills nest in crevices or under protective overhangs near penguin colonies, and time their own breeding to coincide with that of the penguins. They attempt breeding at three years old, but most fail. Once they succeed in breeding, they remain faithful to their territory and their mate, and are strongly territorial during the breeding season. Sheathbills normally lay two or three eggs in a rough nest of tussock, moss, algae, and old bones, built in crevices or under an overhang to protect them from the weather and marauding skuas. Once the chicks hatch in January, parents deliver food by the billful, but do not regurgitate it. When the chicks become independent, at about 50 days, the adults gather in small flocks to forage and roost.

NOT FUSSY EATERS

Snowy sheathbills, the garbage collectors of Antarctica, eat anything small enough for them to swallow, including spilled food, eggs of any other species, excrement, carrion, and mucus from seal's noses. They are restricted to the relatively milder regions around the Antarctic Peninsula, and are omnipresent in penguin colonies.

Snowy sheathbill

CHIONIS ALBA

OTHER NAMES: American sheathbill, Greater sheathbill, Pale-faced sheathbill, Yellow-billed sheathbill
LENGTH: 34–41 cm/13–16 in
WINGSPAN: 75–80 cm/29½ –31½ in
STATUS: about 10,000 breeding pairs
IUCN: not listed   CITES: not listed

With their thick, entirely white plumage and sturdy bodies, Snowy sheathbills are well adapted to the cold Antarctic environment. Their greenish bill sheaths and pink facial caruncles make their faces look untidy. They live and breed around the Antarctic Peninsula and the islands of the Scotia Arc. Many migrate to Tierra del Fuego and Patagonia in winter, but on Signy Island and the Falklands some stay near their breeding areas throughout the winter.

Cormorants

Cormorants are medium to large aquatic birds with webbing across all four toes. They are excellent divers and swimmers but rather awkward on land. Their bodies are heavy and elongated, and they have long necks, heads, and bills. Like pelicans, cormorants have a pouch of loose skin on the throat, called a gular pouch. They use it for holding food and for thermoregulation. On a warm day it is common to see cormorants fluttering their throats to increase the evaporation on the inside of their gular pouch.

Being heavy, cormorants swim quite low in the water, sometimes with just their head and neck raised. They normally stay submerged for only a few seconds while foraging, but they can dive for more than 90 seconds and reach depths of 25–50 meters (65–165 ft). Like all seabirds, they have oil glands to waterproof their feathers, but their feathers also have a special structure that allows their plumage to become waterlogged to adjust their buoyancy for diving. They quickly shake off most of the water when they return to land, but it is common to see cormorants standing on low rocks or on the shoreline with their wings outstretched to dry off their feathers.

Cormorants do not rest on water, so they require dry perches. Because they need to dry their feathers, they are almost always within sight of land, and most do not migrate. Many species are also restricted to relatively shallow water because they forage for small, bottom-dwelling fish and invertebrates. Some species follow coastlines seasonally in search of food, but all the island forms are entirely sedentary.

Most cormorants are very dark brown to black, with a metallic green or blue sheen to the feathers. All the Antarctic and sub-Antarctic shags have some white on their necks and bellies. As the breeding season approaches, mature adults develop white, hairlike plumes and brightly colored facial patches: these are important during courtship, but are lost after breeding. Juveniles are uniformly drab brown with brown eyes, developing the typical adult plumage and eye color as they mature.

Snowy
sheathbill

Tristan da
Cunha
Gough I

0°

Bouvetøya   Prince Edward Is
South
Georgia
Sth Sandwich Is
Is Crozet
Falkland Is   Sth Orkney Is
Is Kerguelen
Heard I
90°W
South Pole
90°E

Balleny Is
Polar Front

Campbell I   Macquarie I
Antipodes Is   Auckland Is
Bounty Is
Chatham Is

Distribution
Breeding sites
180°

Cormorants breed in busy, noisy colonies, often sharing their breeding sites with other species—gannets, boobies, or penguins. Their colonies are on cliffs, low ledges, rocky slopes, or even on sand. Each year, after attracting a female, the male cormorant goes in search of seaweed, mud, or grasses. He flies back with his bill filled with nesting material, and the female plasters it to the nest with mud or excrement. Nests are reused year after year, and many become quite large. Females lay two to four eggs. Both sexes take their turn on the nest, and chicks hatch in laying order after 23 to 25 days. Then the real work begins: with two to four chicks on each nest, all raising their heads and calling loudly for food, the din is unrelenting.

Imperial shag

PHALACROCORAX ATRICEPS
OTHER NAME: Blue-eyed shag
LENGTH: 68–76 cm/27–30 in
WINGSPAN: 124 cm/49 in
STATUS: about 10,000 breeding pairs
IUCN: not listed   CITES: not listed

The Imperial shag is widespread in the southern part of South America, the Falkland Islands, Prince Edward Islands, Iles Crozet, Heard, and Macquarie islands. It has a large white patch on its back, which is visible only in flight. During the breeding season it develops a brilliant cobalt blue eye ring and bright yellow-orange caruncles at the base of its bill. Some populations also develop fine filoplumes that accentuate their upper cheeks or eyebrows.

Antarctic shag

PHALACROCORAX BRANSFIELDENSIS
LENGTH: 75–77 cm/29–30 in
WINGSPAN: 124 cm/49 in
STATUS: 12,000 breeding pairs
IUCN: Not listed   CITES: not listed

Antarctic shags are a little larger than Imperial shags, although similar in appearance. They are found on the Antarctic Peninsula year-round, except during short foraging flights to open water. Living in such a cold climate, they do not hold their wings out to dry and possess an extra-thick coat of down under their contour feathers. They normally congregate in small groups, often on the edges of penguin colonies.

Kerguelen shag

PHALACROCORAX VERRUCOSUS
LENGTH: 65 cm/25½ in
WINGSPAN: 110 cm/43 in
STATUS: 6,000–7,000 birds
IUCN: not listed   CITES: not listed

Smallest of the blue-eyed shags, the Kerguelen shag is restricted to Iles Kerguelen, and is the only shag of the islands of the southern Indian Ocean that has remained a separate species; all others were classified with the Imperial Shag. The Kerguelen shag feeds in shallow water within about 6 kilometers (4 miles) of the coast of Kerguelen. It breeds in small colonies of about 330 nests.

▲ A BUSY COLONY
Imperial shags nest in crowded, sometimes very large, colonies. They lay two to five (usually three) eggs in substantial nests made of grasses, seaweed, and excrement. Once the chicks begin to hatch, these large colonies are a scene of bustling activity with arriving and departing parents trying their best to keep their hungry chicks fed.

Skuas

The skua family, Stercorariidae, is divided into the larger *Catharacta* genus and the smaller *Stercorarius* genus, sometimes known as jaegers. Both groups probably originated in the northern hemisphere, splitting from the gulls about 10 million years ago, but the jaegers have remained wholly Arctic while *Catharacta* skuas are found in high latitudes in both hemispheres.

Skuas have heavier bodies than their gull relatives, and stronger, more hooked bills, reflecting their more predatory lifestyle. Their plumage is predominantly dark, with conspicuous white patches on the top and bottom of their wings, and they employ powerful wing-beats in flight. With strong, hooked claws on their toes, as well as full webbing between them, they are true seabirds of prey. Like the land birds of prey—the raptors and the owls—they have hard, scaly plates called scutes on their legs, and females are larger than males.

Skuas may live for 38 years, and are very faithful to both their breeding territories and their mates. They are migratory during the winter, arriving at breeding areas in early spring, just as the penguins are returning to their colonies. They are mature by age three, but cannot normally claim a territory until they are six or seven. Most pairs produce two eggs. If both eggs survive, the first egg hatches two or three days before the second. The older chick will pick fights with the younger, sometimes killing it but more often preventing it from getting enough food. The younger chick will either die of starvation or stray into other territories in search of food—usually to be killed and eaten by adult skuas. The reproductive success of skuas is very low.

Two *Catharacta* skuas breed in Antarctic or sub-Antarctic regions. Antarctic skuas nest on most of the sub-Antarctic islands, on cold temperate-zone islands south of New Zealand, and on the northern part of the Antarctic Peninsula. On the peninsula they overlap with South polar skuas, which breed exclusively around the coast of Antarctica. Where the species overlap they regularly interbreed, even though Antarctic skuas normally will not allow South polar skuas to nest around penguin colonies, forcing them to forage almost exclusively at sea for fish and krill. Further south, away from the overlap with Antarctic skuas, most South polar skuas nest in loose colonies near penguin or petrel colonies, on which they prey. Outside the breeding season they mainly fish for themselves by plunge-diving at sea.

▲ **SAFE HAVEN**
South polar skuas normally lay two eggs, but accidents caused by aggressive penguins and predation by other skuas often cause pairs to fail before their eggs hatch. The chicks must be protected and brooded regularly, but not constantly, for the first couple of weeks.

Antarctic skua

CATHARACTA ANTARCTICA
OTHER NAME: Brown skua
LENGTH: 63 cm/25 in
WINGSPAN: 120–160 cm/47–63 in
STATUS: 13,000–14,000 breeding pairs
IUC: not listed   CITES: not listed

Antarctic skuas are large, powerful birds with heavy bodies and a fierce demeanor. They are dark brown with yellowish streaks in their neck and nape feathers. Most populations are fairly small, but bigger populations occur on larger islands that harbor more prey species. They are generalist predators who adapt to local conditions, and are also scavengers whose numbers tend to increase around human settlements.

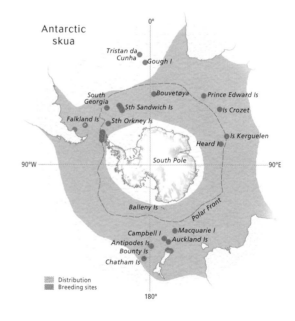

Antarctic skua

Tristan da Cunha
Gough I
South Georgia
Bouvetøya
Prince Edward Is
Sth Sandwich Is
Is Crozet
Falkland Is
Sth Orkney Is
Is Kerguelen
Heard I
South Pole
90°W
90°E
Balleny Is
Polar Front
Campbell I
Macquarie I
Antipodes Is
Auckland Is
Bounty Is
Chatham Is
0°
180°

Distribution
Breeding sites

South polar skua

CATHARACTA MACCORMICKI
OTHER NAME: McCormick's skua
LENGTH: 53 cm/21 in
WINGSPAN: 130–140 cm/51–55 in
STATUS: 5,000–8,000 breeding pairs
IUCN: not listed   CITES: not listed

Smaller than the Antarctic skua and slightly more agile in flight, the South polar skua is the grayest of the *Catharacta* skuas; even so, there is usually a clear contrast between their lighter nape and their darker back. They occur in pale, medium, and dark color forms.

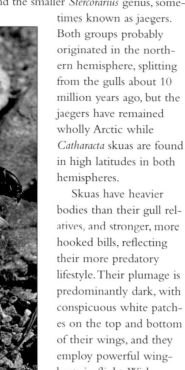

South polar skua

Tristan da Cunha
Gough I
South Georgia
Bouvetøya
Prince Edward Is
Sth Sandwich Is
Is Crozet
Falkland Is
Sth Orkney Is
Is Kerguelen
Heard I
South Pole
90°W
90°E
Peter I Øy
Balleny Is
Polar Front
Campbell I
Macquarie I
Antipodes Is
Auckland Is
Bounty Is
Chatham Is
0°
180°

Distribution
Breeding sites

Gulls and terns

Gulls and terns are found all over the world. They breed in colonies and are territorial: they will defend their own nests, but the entire colony will attack any intruder approaching the colony.

Most gulls live on sea coasts, but some species live well inland and nest near fresh water. Gulls are omnivorous and opportunistic feeders. They eat virtually anything, and are familiar visitors to refuse tips around human habitations, and they can obtain their food in a variety of other ways as well: on the wing, hovering above water, dropping down to pick items off the surface, or landing on the water to surface-feed.

Terns forage by swooping down to pick food from the surface or plummeting into the water.

Kelp gull

LARUS DOMINICANUS

OTHER NAMES: Dominican gull, Southern black-backed gull
LENGTH: 54–65 cm/21–26 in
WINGSPAN: 128–142 cm/50–56 in
STATUS: more than 1,085,000 breeding pairs; 10,000–20,000 in Antarctica and sub-Antarctic regions
IUCN: not listed   CITES: not listed

The only gull species commonly seen in Antarctica, Kelp gulls are most abundant in New Zealand, although they are also found on coasts throughout the southern hemisphere. In Antarctica they occupy the coasts and islands of the Antarctic Peninsula throughout the year; in winter they may move away from the coast in search of open water, but they do not migrate north into South America. They are large, white-headed gulls, with a black mantle and upperwings and completely white tails. Their straight, yellow bill has a red spot at the tip of the lower section.

Antarctic tern

STERNA VITTATA

OTHER NAME: Wreathed tern
LENGTH: 35–40 cm/14–16 in
WINGSPAN: 74–79 cm/29–31 in
STATUS: About 50,000 birds
IUCN: not listed   CITES: not listed

This species, which is the most common tern in Antarctica, nests on most sub-Antarctic islands, as well as along the coasts of the Antarctic Peninsula, New Zealand, and some areas of South America and South Africa. It is smoky gray on its upper parts with a white rump, a black cap on its gray head, and a red bill. Its underparts are mostly white, sometimes with streaks of gray.

Antarctic terns breed in November–December, nesting in colonies in rocky areas near the shore or just inland. Around the Antarctic Peninsula they usually establish their colonies in flat gravel areas away from the beach. If left undisturbed they come back to the same site year after year, but they can be flexible: on Tristan da Cunha, they once nested on sandy beaches, but shifted their colonies to ledges on small offshore outcrops when rats were introduced. During the non-breeding season they move to ice edges to forage; they may migrate northward as far as the South American coast, but they do not migrate into the northern hemisphere. Their favorite prey is small fish and krill, which they obtain by plunge-diving. GM

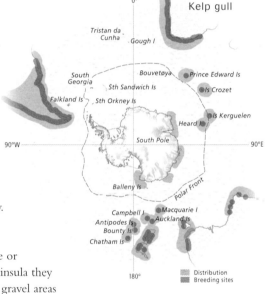

Kelp gull

Tristan da Cunha · Gough I · Bouvetøya · Prince Edward Is · South Georgia · Sth Sandwich Is · Is Crozet · Falkland Is · Sth Orkney Is · Is Kerguelen · Heard I · 90°W · South Pole · 90°E · Balleny Is · Polar Front · Campbell I · Macquarie I · Antipodes Is · Auckland Is · Bounty Is · Chatham Is · 180°
Distribution
Breeding sites

Antarctic tern

Tristan da Cunha · Gough I · Bouvetøya · Prince Edward Is · South Georgia · Sth Sandwich Is · Is Crozet · Falkland Is · Sth Orkney Is · Is Kerguelen · Heard I · I Amsterdam · 90°W · South Pole · 90°E · Balleny Is · Polar Front · Campbell I · Macquarie I · Antipodes Is · Auckland Is · Bounty Is · Chatham Is · 180°
Distribution
Breeding sites

Penguins

With their sway-backed upright posture, smart black-and-white coloration, and endearing waddling gait, penguins are perhaps the most distinctive of birds. Many people think of penguins as strictly Antarctic, but in fact only four species—the Emperor, the Adélie, the Chinstrap, and the Gentoo—breed on the Antarctic Continent proper, while seven others live and breed south of the Polar Front that defines the far southern oceans. Because penguins are flightless and must feed in cool water, they cannot travel across the warm waters of the tropics to inhabit the northern hemisphere. Only a few groups of Galapagos penguins (*Spheniscus mendiculus*) breed north of the equator.

Grouped scientifically into about 17 species, penguins are highly specialized, non-flying, marine birds, ranging in size from the Little penguin, at 1 kilogram (2¼ lb) in weight and 40 centimeters (16 in) in height, to the Emperor penguin, which weighs up to 38 kilograms (84 lb) and reaches 115 centimeters (45 in) in height. All have a dark back and a white front, and each species has a unique pattern, in some cases including orange or yellow, around the head and chest. Possibly the basic dark and pale coloration, repeated with variations, provides both protection from predators and camouflage while hunting: from above, penguins on the ocean surface are dark against the dark water, but

viewed from below, they are light against a light sky.

Male penguins are generally heavier than females and have larger, more powerful bills and longer flippers. In most species the difference between the sexes is hard to distinguish by eye, but behavioral observations do help. Scientists can sex Macaroni penguins, but find distinguishing male and female Gentoo and Chinstrap penguins more difficult, although they have achieved a 95 percent success rate. Adélies are even more difficult to sex by eye, with scientists succeeding only 85 percent of the time. Interestingly, penguins from high and low latitudes are more difficult to sex than mid-latitude species, suggesting there may be some environmental

BALANCING ACT
King penguins do not use nests to care for their eggs, instead they tuck them up on their feet and warm them against a bare patch of skin on their lower belly. Because of this balancing act, a King penguin with an egg will only move a few meters, and even then only if disturbed.

factor contributing to the presence (or absence) of sexual dimorphism. One reason may be the need for males to compete for nesting space and females; larger males would win more battles and so be more successful. A further theory is that if the two sexes are different sizes they can effectively exploit different food resources, but this has never been shown to occur regularly.

Keeping warm—and cool

Unlike most other seabirds, penguins have lost the ability to fly, so they do not need to minimize their weight. On the contrary, they have developed heavy blubber reserves as insulation and for long-term energy storage, an adaptation usually seen only in mammals. Their feathers have also modified from flying equipment into waterproof insulation, thickening the birds up and shaping them to hold air against the body. Penguins may not be able to take to the air, but they do "fly" through water, using their legs as rudders and their rapidly beating wings for propulsion. Swimming compresses their feathers and gradually forces out the insulating layer of air, so that a swimming penguin leaves a fine trail of bubbles.

Polar penguins are well insulated, and can remain in the water almost indefinitely without losing too much heat. But they are designed to retain body heat, and when conditions are too warm they have extreme difficulty cooling down. They shed heat through the undersides of their flippers and through their legs and feet, as well as by fluffing feathers and panting. On a warm day, penguins may lie on their stomachs, flippers raised, feet waving in the air. In the heat of summer, bigger chicks that are still too young to go to sea and dissipate heat into the water have an extremely difficult time.

Fishing the southern oceans

All penguins are carnivores but the various species prefer different prey, ranging from small plankton to large fish and cephalopods. These prey types are all fairly active, so they are difficult to catch and must be swallowed whole before they can escape. Bird beaks do not have

ICY CONDITIONS
The very best nesting sites are those that become snow-free first, allowing for an early start to breeding. However, even the best locations suffer the occasional surprise snow-storm, even in the middle of summer. Very young chicks are not well insulated, but older ones are quite capable of sitting out a snow fall.

▲ HOW FAST?
Despite some enthusiastic
estimates of speeds of up
to 50–60 km per hour
(30–37 mph), small to medium-
sized penguins such as the
Adélies seen here actually
swim at about 5–10 km per
hour (3–6 mph), or up to twice
that in short bursts.

▼ THE WAY HOME
Most penguins seem to navigate by the sun, and are
equipped with an internal
clock to adjust for the time of
day. Experiments have shown
that penguins know which
direction they want to take,
but can find it only by the
position of the sun in the sky.

teeth, but penguin mouths have "teeth" made from keratin. They are small hooks, rather like an extremely rough cat's tongue, that project backwards inside their mouth and prevent their prey from darting to freedom at the last second.

It is estimated that there are at least 23 million breeding pairs of the 11 species of Antarctic penguins living and fishing within the region south of the Polar Front. Scientists have calculated that they must consume at least 539,000 tonnes of squid, 3.6 million tonnes of fish, and 13.9 million tonnes of crustaceans—mostly krill—every year. The great bulk of this fishing takes place near South Georgia and the South Sandwich Islands. This is an immensely productive area, and it supports 24 percent of all the breeding penguins that live within the Polar Front. It is the principal location for penguin consumption of marine resources.

New feathers for old

Each year, penguins undergo a rapid complete molt, in which all their feathers are replaced at once. First the new feathers grow, pushing out the old feathers, which remain attached to the tips of the new. Eventually the old feathers are shed in batches, creating a snowstorm of feathers around the penguin colonies, and then the new feathers grow further and thicken up before the penguin is ready to go to sea again. The process usually takes about three weeks, and during that time every bit of the penguin's energy is focused on renewing its feathers.

Without feathers, penguins have no insulation or waterproofing, so a molting penguin cannot go to sea to feed—it would freeze to death.

From when they begin to molt until their new feathers have grown in, penguins are very exposed to the elements and are at risk of freezing or starving, or both. In order to conserve energy, they move around very little while they are molting; they simply stand on shore sheltering from the weather, and use their blubber reserves to grow new feathers. Even so, the molting phase results in massive weight loss—a molting penguin can lose more than half its body weight in just over three weeks.

Preening, in which birds use their bill to stroke their feathers from base to tip, is essential for penguins. Penguins coat their feathers with an oil secreted by the uropygial gland, located at the base of the tail, which waterproofs the feathers. They squeeze oil out of their uropygial gland with their bill, then run the oil through their feathers. To put it on their head and neck—the hard-to-reach places—they put the oil on the edge of a flipper and rub their head over the flipper area. In some species, partners will preen each other's head and neck; this is called allopreening. While allopreening may help to spread oil, it is considered to be mainly a nurturing behavior. Emperor penguins, which do not seem to have external parasites, have not been observed to allopreen. Adélies, Gentoos, and Chinstraps—the brush-tailed penguins—are also exclusively Antarctic species, and they do not allopreen, but King penguins, which breed in warmer conditions, do. Allopreening is generally more common in warmer conditions, where there are more parasites, but it is thought to be primarily a social behavior that strengthens the pair bond.

Sharing resources

Good breeding locations on the sub-Antarctic islands and along the Antarctic shoreline are few and far between, and during summer, when all penguins must come ashore to reproduce, they gather in very large numbers. During the summer season several species often share a general area, and its food supply, until they can reproduce, molt, and disperse throughout the southern oceans again. To accomplish this sharing of a limited resource, the penguins employ different feeding strategies and target different prey. Usually, one species fishes inshore, very close to land, while other species go farther out to sea to feed. If there are three or more species in the area, offshore feeders specialize in the types of food they hunt, partitioning the prey by both size and type. Foraging depth may be one of the deciding factors in this separation of resources, with larger species of penguins, which are capable of longer, deeper dives, taking larger prey items.

Breeding in colonies

Most penguin species prefer to live in large colonies of thousands, or even hundreds of thousands. To attract breeding partners from among these crowds, penguins put on attention-seeking courtship displays including calling and highly ritualized posturing. Penguins are highly imitative birds, and when one couple begins a courtship ritual, all the neighboring birds tend to copy them. Experiments playing courtship sounds to pairs of Royal and Little penguins found that it increased the

rate of courtship and mating by shortening the time between pairing and egg laying. In further experiments, the cacophony of colony sounds was recorded and played back to the whole colony and the result was an increase in synchronicity of egg laying. This may partially explain why larger penguin colonies tend to be more successful. The acoustic background of a colony influences its seasonal reproductive rate and synchrony, with larger colonies more synchronous than smaller ones, and more successful. Synchronized egg laying produces many chicks all at once, so imitative mating and laying behavior is an effective survival technique, as it floods an area with a large number of chicks all hatched at the same time. So each chick has a better chance of surviving the local predators, and the shorter the period with chicks vulnerable to predators, the fewer chicks can be taken.

Penguins must provide all the protection they can for their eggs. They incubate their eggs in nests of rocks, tail feathers, bones, and sometimes plant material, but these are not good insulators, and do not provide much protection. To protect their eggs and keep them warm,

penguins tuck them into a brood patch, which is a bare patch of skin surrounded by thick feathers low between their legs. During incubation this patch becomes engorged with blood vessels located at the surface of the brood patch. The vessels are heated by the adult's body core and keep the egg warm. When a penguin stands up or swims between bouts of incubating, the stomach feathers close up over the patch and prevent the belly from losing too much heat.

▲ VARYING DIET
Temperatures along the Antarctic Peninsula fluctuate over time and affect the types of prey species available. Adélies have been observed to change their diet to cope with changed prey availability, and may do so regularly.

> FIRST MOLT
By its first winter a King penguin chick (left) can weigh almost as much as an adult. Over the winter, the chick loses weight, but when spring returns and fishing improves, it regains weight. It also loses its stringy chick down, molting into immature feathers, which are similar to adult feathers but not so bright.

Wingless divers

Wingless divers, genus *Aptenodytes*, are the largest living penguin species. There are only two members of this group, the Emperor penguin and the King penguin. Both these stately birds stand at an impressive height of more than 90 centimeters (35 in). The Emperor penguin has a bold splash of golden yellow on the sides of the head, neck, and chest, and the King penguin is similar but more brightly marked with a rich orange color. There is a significant weight difference between the two, with the King weighing only 35 percent of the Emperor.

The two species are quite closely related, but they live in vastly different conditions. Emperors, the most southerly of penguins, live and breed in the harshest of environments, in the fast ice of the Antarctic Continent, whereas King penguins inhabit the comparatively mild sub-Antarctic islands north of the Polar Front.

Neither species builds a nest, even though King penguins live in an area where nest-building materials are readily available. Emperor penguins, on the other hand, raise their young deep in the southern ice, where there is not even a scrap of rock to build a nest.

▲ RUSHING TO MATE
King penguins are usually four years old when they first begin to breed, and after successfully raising a chick, the female is often ready to mate again a month sooner than the male. This means that successful couples don't usually re-mate.

Bringing up baby

King penguins hatch in summer and reach their parents' body weight in about two months, but when winter sets in and prey becomes scarce, they are not regularly fed and so lose weight. The next spring, prey returns, the parents can feed the chicks again, and their weight recovers to about 80 percent of the adult weight before they fledge, at approximately 13 months old. Emperor penguins hatch in mid-winter, four months later than Kings, and mature steadily over five months. They fledge in December with a weight and beak size 40–60 percent that of adults—but with feet 80 percent of adult size. Presumably they need large feet to walk from their birthplace in the fast ice to the open ocean, where they can begin to hunt independently.

Each strategy has its pay-offs. King penguins' two-stage rearing means that energy is not wasted on weak youngsters that do not survive winter, but chicks that do make it through have good survival prospects. The Emperors' shorter rearing period means that they can breed annually and produce more offspring, balancing out the fact that many chicks perish long before reaching the ocean. There is a high mortality rate in Emperor penguin chicks—the cold climate the chicks must weather and harsh conditions that their parents must forage in result in up to 33 percent of each year's eggs and chicks not surviving to fledge.

▶ MUTUAL ADMIRATION
Emperor penguins could not survive the bitter Antarctic winter without cooperation, and the early socialization of chicks is part of this overall strategy for survival. Emperor penguin parents will often shuffle together, face-to-face, so that the growing chicks on their feet can get to know each other.

▶ RECOGNIZING THEIR OWN
Parents returning from fishing expeditions locate their chicks by sound. King chicks are taught to recognize the characteristic rise and fall of their parents' voices from a very young age. The chicks respond with a high-pitched, three-noted whistle.

Emperor penguin

SPECIES: *Aptenodytes forsteri*
LENGTH: 100–130 cm/40–50 in
WEIGHT: 38 kg/84 lb
STATUS: at least 500,000; population thought to be stable
    IUCN: not listed   CITES: not listed

Emperor penguins have blue-gray backs shading to a black tail, and their underparts are white flushed with yellow, strengthening to deeper yellow curving ear patches. Uniquely among penguins, the chicks have beautiful, pale gray body down and a black head with a white eye mask.

Emperor penguins live in 40 known colonies south of latitude 65°, and there are three regional populations, in East Antarctica, the Ross Sea, and the Weddell Sea

area. They spend their lives in the pack ice, avoiding the open water beyond, and breed only on the permanent ice attached to the Southern Continent and nearby islands. Emperors are the only penguins that breed on ice and snow rather than exposed land. In order to feed, they must visit open water, and if open water is not available they dive into the breathing holes maintained by seals and into narrow cracks in the continental ice. They feed on fish, and Emperors have been recorded diving for as long as 18 minutes to depths of more than 400 meters (1,300 ft).

Emperors have many ways of dealing with their rigorously cold environment. Large bodies retain heat more efficiently than small ones, and the Emperor is the largest of all modern penguins. All the Emperor's extremities are reduced in size, limiting the heat that can be lost through

▲ **SURVIVING THE COLD**
Emperors have large, rounded bodies that help them to combat the cold. They are well insulated, mostly by a thick layer of waterproof feathers which overlap like tiles on a roof. A layer of fat under the skin provides further insulation, and doubles as a food reserve.

▶ **CHICKNAPPING**
So strong is the brooding instinct in Emperor penguins that chickless adults will try to steal any chick that leaves its parent for even a few seconds. Up to a dozen adults may try to claim a chick by pushing it under their belly feathers with their pointed bill, often injuring or even killing it in the process. Without a mate to share brooding and feeding, the foster parent must eventually head for the sea to feed, leaving the chick to die.

them. It also has a highly developed blood counter-current mechanism—a heat exchange system designed to retain warmth in the body. The Emperor's body is designed to withstand temperatures of −10°C (14°F) before it must use body energy to keep warm, whereas the King penguin, its warmer-climate relative, must draw on body energy at a temperature of only −5°C (23°F). Like all penguins Emperors have some insulating blubber, but their main form of insulation is their feathers, which are dense and tightly packed, and lock together to trap air in the layer of down against the skin.

Family life

Living as far south as any animal in the world, Emperors must reproduce in indescribably savage conditions. They have little time or energy to spare on courtship, and pair

bonds form quickly. Unlike many penguin species, Emperors do not go fishing to stockpile energy before laying their eggs. After pairing for six weeks, the female lays one egg, and the male immediately takes it from her feet, puts it on his own feet, and covers it with his brood patch to keep it warm. In this position the egg is sheltered by a fold of skin, and the male rocks back onto his heels to minimize the amount of ice touching his feet. Meanwhile, the female has done her part for the time being, and immediately heads for the sea to fish.

The males incubate the eggs throughout the winter in constant darkness, surviving by huddling together; the temperature often drops to −60°C (−76°F), with winds up to 180 kilometers per hour (110 mph). After nine weeks the chicks hatch and the males give them a small feed. Shortly afterwards the females return and start persuading the males to give up their charges; eventually the males hand the chicks over to their mates.

Now, at last, the males are free to make for the sea to fish. They will have been fasting for four to six months and will have lost about half their body weight by the time the females return to claim their offspring, but they still usually have to walk about 100 kilometers (60 miles) across the ice to reach open water and a much-needed meal. Males never die of starvation while caring for their chicks. If the female is late in returning, the male simply abandons the egg or the chick and makes for the sea to feed before it is too late for him to survive.

After the chicks have begun to hatch, two lines of Emperor penguins snake between the colony and the sea, one coming and one going. The chicks need constant feeding, and the parents take turns providing meals for about seven weeks. After that, the growing chick needs more food than a single parent can provide, so the chicks huddle together for warmth while both parents go to sea at the same time to fish.

▲ SURVIVAL RATES

Life as an Emperor penguin chick is very harsh, and many do not survive. Only 65 percent become independent from their parents, and of those that do fledge, less than 20 percent manage to survive their first year. If they make it through a few years, their survival rate improves dramatically to 95 percent in adulthood—and then they will probably live for another 20 years.

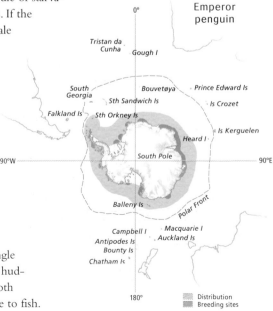

Emperor penguin

0°

Tristan da Cunha
Gough I

South Georgia
Bouvetøya
Prince Edward Is
Sth Sandwich Is
Is Crozet
Falkland Is
Sth Orkney Is

Is Kerguelen

Heard I

90°W
South Pole
90°E

Balleny Is
Polar Front

Campbell I
Macquarie I
Antipodes Is
Auckland Is
Bounty Is
Chatham Is

180°
Distribution
Breeding sites

Brush-tailed penguins

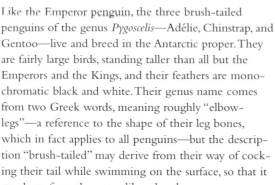

Like the Emperor penguin, the three brush-tailed penguins of the genus *Pygoscelis*—Adélie, Chinstrap, and Gentoo—live and breed in the Antarctic proper. They are fairly large birds, standing taller than all but the Emperors and the Kings, and their feathers are mono-chromatic black and white. Their genus name comes from two Greek words, meaning roughly "elbow-legs"—a reference to the shape of their leg bones, which in fact applies to all penguins—but the description "brush-tailed" may derive from their way of cock-ing their tail while swimming on the surface, so that it stands up from the water like a brush.

All of these penguins nest on the land around the Antarctic Continent and its islands, colonizing headlands and hills exposed as the snow retreats with the coming of summer. Exposed land is very limited, but is essential to the success of a colony. The penguins build their nests from small rocks uncovered as the snow melts. It is not much of a nest, but is better than nothing: meltwater drains away below it, and the rocks provide shelter from the wind. To shape the nest, one partner lies down and kicks backwards with one foot, scraping out a cup-shaped hollow among the stones. In fact, brush-tailed penguins value stones so much that when they return from the sea they often bring one to their partner.

The females lay two eggs and the parents incubate them in turns, lying along their bowl-shaped rocky nest with the eggs tucked into their insulating brood pouch. Given good weather and good fishing both chicks will fledge, but normally only one survives.

▲ **CHANGING POSITION**
To receive regurgitated food from its parent, a very young chick must stretch upward, usually thrusting its whole head into the parent's beak. Older chicks often just open their beaks while a parent tips a load of food in.

Adélie penguin

SPECIES: *Pygoscelis adéliae*
LENGTH: 71 cm/29 in
WEIGHT: 5 kg/11 lb
STATUS: at least 2.6 million pairs;
    10 million immature animals
    IUCN: not listed   CITES: not listed

Adélie penguins are everybody's idea of the prototypical penguin. They have a clearly demarcated and simple but strong black-and-white coloring, with a blue-black back and head and a white chest and throat. Their sole touch of color is a small patch of dark orange on the base of their bill, which is partially feathered, the black feathers making it look shorter than it really is. Around their eyes Adélies have vivid white rings, which the young develop long before they grow their adult plumage. This most arresting feature is a vital part of the penguin's armory; eyes down and crest erect, it exposes even more white in the sclerae of its eyes while it performs its three main displays of aggression: the direct stare, the fixed and alternate one-sided stare, and the crouch. The startling contrast of black and white constitutes an unmistakable threat.

Adélies are rarely found north of 60°S, and always remain entirely south of the Polar Front, living on the Antarctic Continent and nearby islands. They prefer to live within the pack ice, but small numbers colonize the South Shetland, South Orkney, and South Sandwich islands. Adélie rookeries will form anywhere that the penguins can reach, as long as it is free of ice. Sometimes the route to the rookery is very steep, but this does not deter the penguins as long as there are no large steps, which the birds cannot negotiate with their short legs.

Adélie colonies may be very large, with populations of 20,000 to 30,000 birds com-mon. There are seldom more than 100,000 penguins in one area, but occasionally a single large colony may be home to more than one million birds.

Adélie penguins are shallow divers, and like all the other penguin species they feed by pursuit diving, pecking out their food as they swerve from side to side under water. Adélies prefer

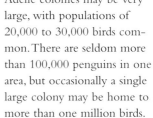

◄ **ANCIENT SITES**
Some penguin rookeries have been in the same location for thousands of years, especially those on the Antarctic Continent. Rookeries like this one on the Windmill Islands are about 3,300 years old, but some Ross Sea rookeries have been oper-ating for 13,000 years. In contrast, the Peninsula's oldest known colony is 644 years old.

to eat euphausiids, but will also consume fish, amphipods, and cephalopods. Their diet varies according to what is available and where they are. Evidence suggests that Adélie penguins breed most successfully when krill is locally abundant.

The gathering of the clans

Adélie penguins are gregarious, and often gather together to molt on floes and bergs in the pack ice. Large gatherings can often be found on the lee side of a hummock or pressure ridge, sheltering until they can take to the water and fish again. Before going to sea, they congregate at the water's edge, calling back and forth to birds already in the water, until finally one penguin jumps. Many of the foremost birds follow this lead immediately, and then the birds at the back of the group move to the front and start calling to those already in the water—perhaps as a protective device to limit the risk of being taken by Leopard seals. When an Adélie comes from the sea in to land, it surfaces approximately 20 meters (65 ft) from the beach, looks around, then swims in and launches itself 2 meters (6 ft) out of the water, gripping the ice or rock with its long toenails to prevent itself from falling back in.

At the onset of winter Adélies disperse northward from the pack ice, the younger birds leaving the colonies first. Many depart before molting, moving into the ice, close to krill supplies, and sometimes returning to their breeding sites when storms open pathways through the ice. But the birds usually spend autumn and early winter in pack ice up to 650 kilometers (400 miles) north of the Antarctic Continent, and juveniles sometimes winter even further north than adults.

Homeward bound

In spring Adélies go back to their breeding colonies. The first to arrive are the older, more successful individuals, up to eight years old; the last are the two-year-olds, who are about to breed for the first time. But everywhere the time of home-coming is governed by the ice, and first-time breeders may be very late when the pack ice is heavier and more persistent than usual. Males arrive about four days before females, and the middle of the colony houses the more successful breeders, generally older, more mature couples. Here reproduction is usually highly synchronized, with successful pairs returning to rebuild their nests in exactly the same location year after year.

Adélies cannot lay eggs until their colony surfaces are clear of snow, so once it has found a suitable location a colony stays put. After the female has laid, the males take the first incubation shift of two weeks and then the parents swap. Either parent may be on the nest when the eggs hatch. After hatching, the parents usually brood alternately, taking two-day shifts each for three weeks until the chick is large enough to be left on its own. Then both parents must go to sea in order to satisfy their offspring's voracious appetite.

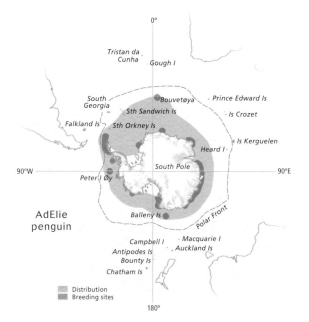

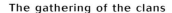

Tristan da Cunha
Gough I
South Georgia
Bouvetøya
Prince Edward Is
Sth Sandwich Is
Is Crozet
Falkland Is
Sth Orkney Is
Is Kerguelen
Heard I
90°W
South Pole
90°E
Peter I Øy
AdElie penguin
Balleny Is
Polar Front
Campbell I
Macquarie I
Antipodes Is
Auckland Is
Bounty Is
Chatham Is
Distribution
Breeding sites
180°

Chinstrap penguin

SPECIES: Pygoscelis antarctica

COMMON NAMES: Ringed, Bearded, Antarctic,
  or Stone-cracker penguin

LENGTH: 72 cm/30 in

WEIGHT: 3.8 kg/8½ lb

STATUS: 6.5–7.5 million pairs;
  some populations are increasing rapidly
  IUCN: not listed   CITES: not listed

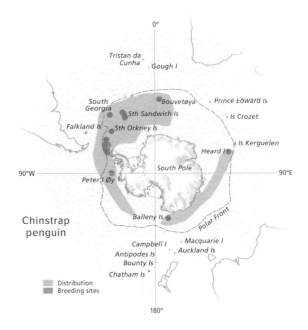

Chinstrap
penguin

Distribution
Breeding sites

Chinstrap penguins have a white throat, chest, and
under parts, and a black body and head. Their bill is
black, and a distinctive thin black line runs from ear
to ear through the white of their throat. It is this band
that gives them their main common name, Chinstrap,
but these penguins were frequently seen by early
Antarctic explorers, and were also named for their
habits (hence "Stone-cracker").

There are eight regional populations of Chinstrap
penguins, all based inside the Polar Front; 99 percent of
all Chinstraps live within the southern Atlantic Ocean.
They inhabit both the sub-Antarctic and the Antarctic
proper, preferring light pack ice. They breed on ice-free
land on the Antarctic Peninsula and the islands south
of the Polar Front, largely within the Scotia Arc.
Chinstraps are very agile climbers, and their favorite
nesting sites seem to be rocky slopes, headlands, rough
foreshores, and high cliff edges.

These penguins are specialist feeders and eat mostly
crustaceans, especially krill. Chinstraps dive to less than
50 meters (160 ft) most of the time, spending 40 percent
of their time almost on the surface at a depth of less
than 10 meters (32 ft), and 90 percent of their time at
less than 40 meters (130 ft). Their maximum dive depth
seems to be 70 meters (230 ft), but they probably reach

▼ MIGRATION

The Antarctic Peninsula is
currently warming, and this
seems to be one of the
factors that is encouraging
Chinstraps, like those shown
here, to extend their range
further south. Colonies found
at the southern end of their
Peninsula distribution limit did
not exist 50 years ago.

Studying stress

The study of penguins is fascinating, but the disturbance of breeding penguins by human activity is the subject of constant concern and research. Initial investigations looked at obvious signs of distress, such as fleeing the nest and abandoning young at the approach of humans, or at the success or failure of reproduction. Currently, less obvious marks of distress, such as increased heart rates, are being examined and compared with the effects of normal stresses—for example, the disturbance shown by a nesting bird at the approach of a skua or other scavenger—but the effects and their implications are difficult to separate from normal fluctuations in reproductive success. One ingenious new way of measuring stress is a heart-rate monitor in the shape of a fake penguin egg, which is placed under experimental penguins.

this depth only when they are desperately trying to find food for their demanding chicks. Chinstrap dives have been timed; they generally feed for about two and a half hours, covering only 5 kilometers (3 miles) over a five-hour trip. Their shallow feeding dives last only about one and a half minutes each, with just over 30 seconds between dives. One reason for the current research interest in Chinstrap eating habits is that these penguins seem to have been very successful recently. Their range has expanded and their numbers are rising, perhaps because there is more krill in the southern oceans with the reduced number of krill-eating whales, or possibly because of the reduction in winter ice in the waters where Chinstraps hunt.

Noisy breeders
Chinstraps are extremely noisy, and it is surprising just how cacophonous a colony can be. All their calls are louder than those of the other brush-tailed species, and they also seem to call more frequently than other penguins. Away from the nest, Chinstraps are very

gregarious and very inquisitive. Of all penguin species, they are the least likely to back away from an encounter and the most renowned for attacking humans who encroach on their territory—hence their reputation as feisty and aggressive.

While they are breeding Chinstraps stay close to their home colonies, traveling a maximum of 100 kilometers (60 miles) in the chick-rearing phase, and perhaps slightly further when the eggs are incubating. During this period they concentrate their hunting energies around nearby ice floes, presumably in search of the krill that is likely to be found nearby. After reproduction, when Chinstraps molt, some choose to wait it out just inland of their colony, while others seem to prefer to molt at a greater distance.

Retreating from the ice
In winter, as the temperature falls and the pack ice starts to expand northwards, Chinstraps move into the open waters north of the pack ice. They leave the Antarctic Peninsula and nearby King George Island in early April and do not return until spring (late October). But they stay together, and over the winter large groups congregate in the open water north of the pack ice. Little is known about where Chinstraps spend their non-breeding season, but they probably attempt to follow the krill and other organisms on which they feed. Chinstraps have been spotted a staggering 3,200 kilometers (2,000 miles) from their breeding site, but this figure may be based on individual birds that are extending the range of the species, rather than on the movements of the main population.

⋏ CHASING FOOD
Chinstrap parents returning with food often lead older chicks on a chase through the colony, making them scramble after them for some time before finally stopping to feed them. This behavior possibly serves to separate weaker chicks from stronger ones when there is only enough food for one chick.

➤ SPECIALIZED FEATHERS
Penguins' feathers are highly specialized for insulation. The very short, broad, and flat rachis (center shaft) reduces the overall length of the feather and allows it to lie flat against its neighbor, and the closely spaced barbs, especially near the rachis and tips, provide good interlocking. There are also small, downy afterfeathers, which form a second layer of insulation.

Gentoo penguin

SPECIES: Pygoscelis papua

COMMON NAME: Johnny penguin

LENGTH: 75 cm/30 in

WEIGHT: 5.5 kg/ 12 lb

STATUS: more than 300,000 pairs

   IUCN: Lower Risk–near threatened

   CITES: not listed

> SETTING UP

Gentoos will nest with Adélies or Chinstraps, but generally prefer lower ground than the other two brush-tailed species when nesting in the same area. Not surprisingly, they prefer to form colonies away from elephant seals and flood zones, and on the Peninsula usually site their colonies less than 100 meters (330 ft) from the shore.

▼ BEING RESOURCEFUL

Suitable building materials for nest building are often very scarce, so supplies will be obtained from any source. If small stones are not available, Gentoos may utilize coal, small bits of wood, and other debris found near some Peninsula stations.

Gentoos have a triangular white flash above each eye, white flecks on the crown of their head, and a bright orange bill and feet. They live slightly further north than other brush-tailed species, remaining close to their breeding islands all year in sub-Antarctic and Antarctic regions right around the South Pole, but they move no further south than 65°S, shunning the most southerly pack ice and the Southern Continent itself, except near the Antarctic Peninsula. Most breed within the Southern Ocean, but about 25 percent breed north of the Polar Front. Taxonomists identify two Gentoo subspecies: the southern group, *P. papua ellsworthii*, is fully migratory, following a clear path each year, and has smaller feet, flippers, and bill than the northern *P. papua papua*, which disperses from its breeding areas but does not migrate in the true sense.

When they are not breeding, Gentoos mostly stay around the winter pack ice. The further south they breed during summer, the more likely it is that the whole colony will move northward for the winter, but generally they remain near their colonies, some staying as far south as possible in nearby open water. Many leave their breeding site after molting to feed and regain weight, but they never totally abandon a colony, and populations build up again after a few months. Nevertheless, whole colonies of Gentoos have been known to move on from time to time, perhaps to escape parasite infestation: large numbers of ticks have been observed on Gentoos. They peck constantly at each other and often squabble with their Gentoo neighbors over nesting materials, but they do not come to blows with other penguin species, even when competing species are nesting close by.

Finding food

Most Gentoos feed during the day and return to their colony in the evening. They generally feed inshore, finding shallow-water prey when breeding. Early in the day, they go to sea *en masse*; later they take to the water individually. They tumble in and swim briefly, then surface and clean themselves of the accumulated muck of the colony by rolling about and wiping their bodies all over with their flippers. When they come in to land after fishing, they will pause to inspect the landing site, and then surf in on a wave. They leap onto ice or rocks, then shake their head and preen before moving on to their nest. Gentoos living near Antarctica mostly consume euphausiids, but those that breed north of the Polar Front eat more fish. If need be, they will dive deeper than 165 meters (530 ft) in search of their prey, but their diving is often less than 20 meters (65 ft).

Breeding behavior

Gentoos breed on ice-free land on the sub-Antarctic and Antarctic islands, and the Antarctic Peninsula, and may form colonies some distance inland or directly on the coast. They have been observed to occupy slopes, flats, terraces, valleys, headlands, ridges, and cliff tops, but some scientists have suggested that underwater land-forms dictate these penguins' breeding sites, so that they can always breed close to the shallow waters that are their most profitable hunting areas.

Around 90 percent of Gentoo penguin couples locate their former mate in the next breeding season, but, unlike the Adélie, Gentoos often move to new nest sites, presumably because they have a longer breeding season, so it pays them to wait for a successful partner to come back before beginning to breed. When Gentoos pair at the start of the breeding season, the male collects stones, moss, and dirt, depending on what is available, and starts to build the nest. After pairing, the male devotes his attention to foraging for nesting materials and leaves the actual building to his mate. Once the eggs are laid, individual incubation shifts last from about 24 hours to four days. At the northern end of their range, Gentoos start to lay in winter or very early spring, but on South Georgia they do not lay until the end of October, and on the Antarctic Peninsula they

may not lay until mid-November or later. These times may vary depending on snow cover, with the breeding season being delayed if there is heavy snow, and possibly getting under way early if the seasonal snowfall is unusually light.

Gentoos mostly choose elevated nesting sites at a safe distance from the breeding grounds of the elephant seals that share their regions. Most lay two eggs. In good years both chicks survive, but this is rare, except on the Peninsula, where usually both chicks survive or neither. Survival rates for chicks can almost always be traced directly to food supplies during the breeding period. Gentoos feed very close to their nest sites, and eat about every 24 hours—much more frequently than the other brush-tailed penguin species. This daily feeding during mating, laying, and incubation seems to result in Gentoos being less likely than other brush-tailed penguins to abandon their chicks.

Young Gentoos first venture from the nest at about 20 days of age, and by 29 days most have joined other youngsters in a crèche, leaving both parents free to go to sea to fish. The youngsters fledge and are ready to head out to sea at around 80–100 days, but at first they often spend more time on the beach than in the water.

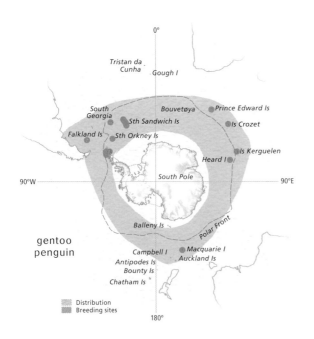

gentoo penguin

Distribution
Breeding sites

What's in a name?

Both the common and the scientific names of the Gentoo penguin are mysteries. Gentoo, its common name, means "pagan" in Hindustani—a word believed to be of Anglo-Indian/Portuguese origin and somehow related to stories about dancing girls, but its etymology is unclear. As for the Gentoo's scientific name, *papua*, the first stuffed specimens to reach England for describing were thought to be from the island of New Guinea, and were named *papua*, which was the Malay word for "curly." Nothing about this penguin merits the description "curly," so perhaps it is simply a reference to the geography of its supposed place of origin.

True divers

Penguins of the genus *Eudyptes* are sometimes called
"crested" penguins, a reference to their jaunty yellow
"eyebrow" plumes. They occupy various habitats in the
southern hemisphere, though none is truly Antarctic,
and some live well within the sub-Antarctic region.
They range from 45 centimeters to 70 centimeters
(18–28 in) in height, and all have red or red-brown eyes.
All species except the Fiordland penguin spend up to
five months of the year at sea.

Eudyptes penguins differ from other penguin species
in several ways. Allopreening, where one penguin preens
another, is common in this group of penguins, although
it is rare in most other groups. Some Eudyptids live in
small groups, or even in single pairs, hidden away from
their neighbors, whereas other penguin species tend to
be extremely gregarious—Rockhoppers often share
their rugged rookeries with albatrosses and shags. They
always lay two eggs; the second, laid about four days
after the first, is as much as 70 percent heavier than the
first, which is almost invariably discarded.

Macaroni penguin

SPECIES: *Eudyptes chrysolophus*
LENGTH: 71 cm/30 in
WEIGHT: 5–6 kg/11–13 lb
STATUS: about 12 million pairs; some
    populations increasing, some in decline
    IUCN: Vulnerable   CITES: not listed

Macaroni penguins are the largest of the
Eudyptes. They are black and white, with
an orange bill and feet, and a bright
crest. The crest is dark yellow or orange,
growing back along the brow line from
the center of the bird's forehead, droop-
ing over the edge behind the eyes, and
hanging rather long at the back of the
head. These birds were named for their
resemblance to London dandies who

▲ ROCKHOPPING
The rockhopper penguin is the smallest of the
true divers. True to their name, rockhoppers
nest among ledges, rocks and grass, and prefer
to come ashore amidst large rocks and rough
seas. On arrival, they scramble from the sea,
hopping ahead of the incoming waves with
both feet together. This preference for hopping
is not shared by other members of the genus.

◄ INTER-SPECIES MATING
Macaronis are the most common penguins in the world, but even with around 6 million Macaronis of the opposite sex to pick from, some birds choose to mate with penguins from other species, including Rockhoppers.

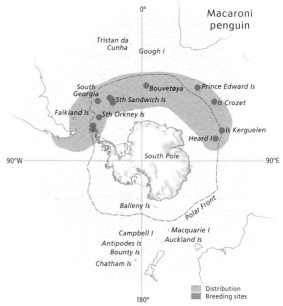

went in for foppish dress and hairstyles associated with Italian culture.

Most Macaronis live south of the Polar Front, in sub-Antarctic and warmer Antarctic waters. They occupy colonies along the Scotia Arc and on South Georgia, and a few pairs nest with other species in the South Sandwich Islands. They often drift for long distances north and south of their colonies, but always remain north of the pack ice.

There are about 12 million breeding Macaroni pairs within the Southern Ocean, making up more than 50 percent of all breeding penguins south of the Front. This is the dominant penguin in the Southern Ocean, in terms of both biomass and consumption of resources, accounting for millions of tonnes of crustaceans and fish from the Southern Ocean each year.

Macaronis are noisy, pugnacious birds. They like to establish their rookeries on gently sloping ground, not too close to the water, with a pathway up along smooth, easily traveled stream beds. Small colonies often share rookeries with Chinstraps or Gentoos.

Breeding Macaronis reach their colony between September and November. The older birds come ashore earlier than the younger ones, and stay longer. The males arrive first and set about selecting or repairing a nest site, and the females come in eight days later. Then the birds pair, laying their eggs between October and November. Like the other *Eudyptes* penguins, Macaronis lay a very small first egg, which is usually lost after the second egg is laid four to six days later. The female takes the first incubation shift and the male takes the second, both about 12–14 days long. Both parents preside over hatching, which takes place some time between mid-November and early January. The male broods the chick for the first two to three weeks while the female makes several hunting trips to the sea to feed herself and the chick. The chicks fledge after about 65 days, and then both parents go to sea to feed for a few weeks, returning to the beaches in March to molt.

Royal penguin

SPECIES: Eudyptes schlegeli
LENGTH: 70 cm/28 in
WEIGHT: 5–6 kg/11–13 lb
STATUS: 850,000 pairs
  IUCN: Vulnerable   CITES: not listed

Many scientists believe that the Royal penguin is a sub-species of the Macaroni, the only difference being the coloration of the face, chin, and throat. Royals have a white or pale gray face, and there may be some black in the throat, but they have the massive head plumes and the very large bill characteristic of Macaronis. Royals are largely, but not entirely, limited to Macquarie Island, south of Australia. They are big for crested penguins, and extremely robust. Like most penguins, they walk by alternating their feet, but sometimes they hop along with both feet together, like their Rockhopper relatives.

Even though both sexes have very large bills, Royals are one of the easiest penguin species to sex when seen in pairs: the males have larger bills, and the females are seldom as white-cheeked as the males—more often gray.

Whereas most penguins are breeding by three to six years of age, it may be 11 years before Royals breed. They breed in the same pattern as most *Eudyptes* penguins, in areas where vegetation is available. They excavate nests from sand or stones, and carry stones to their nesting site. In sandy areas, they also scoop sand into their bill and bring it to the nest site.

When they are not breeding, Royals either migrate or disperse to sea, where they spend much of their time, but little is known about their sea-going behavior or their sea range.   LW

Part IV

ANTARCTIC EXPLORATION

It was not until the fifteenth century AD that the world began to gain accurate knowledge of the area beyond the Arctic Circle—and the Antarctic remained virtually unexplored until early in the twentieth century. Explorers of the eighteenth and nineteenth centuries sailed in frail wooden vessels into the harshest environment on earth, and gradually a true picture of the southern polar region emerged. National interests moved in and through heroic effort most of Antarctica was charted—though by no means conquered.

Early Explorers

1487–1900

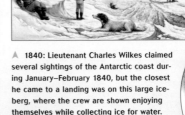

▲ 1830–39: James Weddell reported that at the time of his visit to South Georgia the "sea-elephants were nearly extinct ...."

▲ 1840: Lieutenant Charles Wilkes claimed several sightings of the Antarctic coast during January–February 1840, but the closest he came to a landing was on this large iceberg, where the crew are shown enjoying themselves while collecting ice for water.

▶ 1840: Wilkes (1798–1877) led the first United States expedition to Antarctica. Six ships set off in 1838, but only two survived the whole expedition, and none reached the continent, although lookouts on three ships reported sightings.

◀ 1837–40: Dumont d'Urville's voyage to the South Pole was his third major expedition for the French Navy. It was on this voyage that he made his most famous Antarctic discoveries, which earned him approbation and promotion to rear-admiral.

▲ 1570: This map of the Americas by Ortelius was published in 1570, eight years before Sir Francis Drake proved there was a sea route south of Tierra del Fuego. It shows Magellan Strait as separating South America from *Terra Australis*.

▶ 1837–40: Dumont d'Urville's ships, *Astrolabe* and *Zélée*, were naval corvettes, and not suited to pushing through heavy pack ice. In February 1838 both vessels were beset by ice for five weary days.

| Aristotle suggests that the landmass of the northern hemisphere must be balanced by a large landmass in the south that later became known as *Terra Australis Incognita*, "The Unknown South Land." | Explorers, including Vasco da Gama, Pedro Cabral, Amerigo Vespucci and Vasco Núñez de Balboa, push progressively further into the southern oceans. | Sir Frances Drake leads the first British expedition to circumnavigate the globe, and accidentally discovers where the Atlantic and Pacific oceans meet in what is now Drake Passage. | The first documented sightings of the Antarctic Peninsula are made by Bransfield and Smith (in January), and Nathaniel Palmer (in November). | Thaddeus von Bellingshausen sails further south than Cook, and on 31 October completes his circumnavigation of the South Pole. He discovers and names Peter I Øy and Alexander I Land, and maps the South Shetland Islands. |

5th Cen. BC	4th Cen. BC	1487	1497–1513	1519–22	1577–78	1773	1820	1821	1820–21

| The Greek philosopher Parmenides divides the world into five climatic zones and suggests the zones at the poles are frigid. | Bartolomeu Diaz launches the age of Antarctic exploration when he sails down the west coast of Africa. | Ferdinand Magellan circumnavigates the globe, sailing through what is later named the Strait of Magellan between the island of Tierra del Fuego and the South American continent. | James Cook and the crew of HMS *Resolution* and *Adventure* become the first men to cross the Antarctic Circle. During the summers of 1772–75 they circumnavigate Antarctica, crossing the Antarctic Circle four times. | The crew of an expedition skippered by John Davis are probably the first men to set foot on the Antarctic Continent. |

Background image: As soon as he sighted the Ross Ice Shelf, which he named the Victoria Barrier, James Clark Ross realized that "we might with equal chance of success try to sail through the cliffs of Dover, as to penetrate such a mass." He described these icy white cliffs as "extending from its western extreme point as far as the eye could discern to the eastward."

1839–43: James Clark Ross had already made six Arctic expeditions, and had located the North Magnetic Pole, when he sailed south in 1839 to find its southern equivalent. His was the first British naval voyage to Antarctica.

1839–43: To avoid a huge iceberg, which emerged out of the gloom in a storm, *Erebus* turned, only to find *Terror* "running down upon us, under her top-sails and foresail." *Terror* ploughed straight into *Erebus* and "the concussion when she struck us was such as to throw almost everyone off his feet" wrote Ross later.

1898: The scientific expedition led by young Belgian naval lieutenant Adrien de Gerlache was the first to face the rigors of an Antarctic winter.

1895: Henryk Bull's *The Cruise of the Antarctic* fired enthusiasm for commercial whaling in the Ross Sea region and also sparked off the Heroic Age of Antarctic exploration.

THE CRUISE OF THE "ANTARCTIC"

1898: Watched by a curious Emperor penguin, *Belgica* is shown beset by ice in the Bellingshausen Sea in early March 1898. At first a novelty to the crew, the icy scene became monotonously familiar over the ensuing 377 days until the ship was freed.

1899: After landing at Cape Adare with Henryk Bull in 1895, Carsten Borchgrevink returned there in February 1899 as the leader of his own British Antarctic Expedition, determined to winter on the continent.

1898: Polish geologist Henryk Arçtowski maintained his program of observations throughout the *Belgica* expedition's voyage and, despite poor conditions, made some significant discoveries.

1899: Physicist Louis Bernacchi carried out a full program of work, even when ice-bound. He returned to Antarctica with Scott's *Discovery* expedition in 1901–04.

Sealers and whalers exploit the Antarctic waters.

The sealer and whaler John Balleny carries out the first landing below the Antarctic Circle, on 12 February.

James Clark Ross and his ships *Erebus* and *Terror* explore Antarctica. They take the first Antarctic sea soundings, discover the Ross Sea, and establish that the South Magnetic Pole was far inland, behind a high mountain range.

Henryk Johan Bull, Carsten Borchgrevink, Leonard Kristensen and crewman von Tunzelman make the first recorded landing on the Antarctic continent at Cape Adare in January 1895. All except Bull subsequently claim to have been first ashore.

Carsten Borchgrevink leads a British Antarctic Expedition in the *Southern Cross*. As planned, they land at Cape Adare in Victoria Land and become the first men to spend a winter on the Antarctic continent.

| 1823 | 1830–39 | 1837–40 | 1839 | 1840 | 1839–43 | 1872–76 | 1895 | 1898–99 | 1899 |

James Weddell sails furthest south and discovers what is now the Weddell Sea.

Jules Sébastien César Dumont D'Urville commands an expedition in the *Astrolabe* and *Zelée*. He discovers and names Adélie Land after his wife.

Charles Wilkes leads the United States Exploring Expedition. Wilkes sees the Antarctic coast several times during January and February 1840, but the closest he comes to a landing is on a large iceberg.

HMS *Challenger* collects a wealth of scientific data on an extensive expedition. On 16 February 1872 it crosses the Antarctic Circle, the first steamship to do so.

The scientific expedition in the *Belgica*, led by Adrien de Gerlache, is beset in the Antarctic Peninsula pack ice for a year and 12 days. The party is the first to endure an Antarctic winter, and collects meteorological readings throughout its enforced stay.

First speculations

Antarctica is the only continent that, from the perspective of human thought, began as a sophisticated concept emerging from a series of deductions. In the sixth century BC Greek philosopher and mathematician Pythagoras calculated that the earth was round, and about a century later Parmenides divided the world into five climatic zones not unlike those that we know empirically today. He postulated frigid zones at the poles, a torrid zone at the equator, and temperate zones separating these uninhabitable extremes of heat and cold. In the fourth century BC Aristotle suggested that the landmass of the northern hemisphere must be balanced by a large landmass in the south; later this became known as *Terra Australis Incognita*, "The Unknown South Land." Aristotle gave it a name: at the time, the north lay under the constellation Arktos (the Bear), so he called the other end of the world Antarktikos ("opposite to the

⌃ MISLEADING IDEAS
Vandenburg Glacier, East Antarctica. Claudius Ptolemy's assertion that the unknown southern land was fertile and inhabited, and was connected to the continents already known to exist, caused great confusion for the explorers Christopher Columbus and James Cook.

north"). Aristarchus, a third-century astronomer, was the first to expound that the earth rotated around the sun.

In 240 BC, Eratosthenes of Cyrene (now Egypt's Aswan)—who coined the word "geographica"—calculated the earth's circumference by comparing shadow angles in two distant locations at the summer solstice. It was a simple, ingenious, and remarkably accurate method. The units he used were imprecise, but even so, he overstated the circumference by only about 15 percent. By AD 200 philosophers such as Pomponius Mela had postulated the existence of a cold continent at the southern pole of a globe roughly the size that we now know it to be, spinning around the sun. It was the nearest to the truth that anyone would be for 1,500 years.

Claudius Ptolemy, the influential Alexandrian geographer of the second century AD, added greatly to cartography with his refinement of grid lines of latitude

and longitude to give every location fixed coordinates, northern orientation for maps, and various scales for different uses. However, his mistakes had an equally great effect. He agreed that there must be a southern land, but thought it linked the known continents and was fertile and inhabited. He also ignored the work of Eratosthenes, preferring that of the Greek astronomer Poseidonius, whose theories yielded a calculation of the earth at about 75 percent of its true size. Because of Ptolemy's errors, Christopher Columbus expected to find Japan where he encountered America, and James Cook set out to find the riches of the supposedly fertile Great South Land.

The dark ages

But after Ptolemy and long before Columbus and Cook, an age of ignorance set in. In AD 391 the fabulous libraries of Alexandria were destroyed and an era of dogma based on literal interpretation of the Bible began. Theological objections to the concept of an inhabited southern land proliferated. The Old Testament *Book of Isaiah* states: "It is He that sitteth upon the circle of the earth … that stretcheth out the heavens as a curtain, and spreadeth them out as a tent to dwell in." It was an utterance that gave rise to a perception of the earth as a flat disk. The New Testament *Gospel According to Saint Mark* records God's command to Christ's disciples: "Go ye into all the world, and preach the gospel to every creature." The disciples did not go to the South Land—therefore it could not exist.

The age of exploration

When the Christian city of Constantinople (now Istanbul) was occupied by the Turks at the end of the fourteenth century, fleeing scholars brought Ptolemy's works back to Italy. The *Geography* was translated into Latin in 1405 and hundreds of copies were printed by the end of the century. Europe started to look outwards. The Portuguese Prince Henry the Navigator (1394–1460) is regarded as the initiator of the great age of exploration; although he himself did not travel much, he was the patron of many voyages in Portuguese caravels that ventured far down the west coast of Africa. Henry wanted to reach India by sea for missionary and trade purposes, but the furthest his vessels sailed was to the coast of Sierra Leone. It was left to his compatriot Bartolomeu Diaz (c.1450–1500) to launch the age of Antarctic exploration.

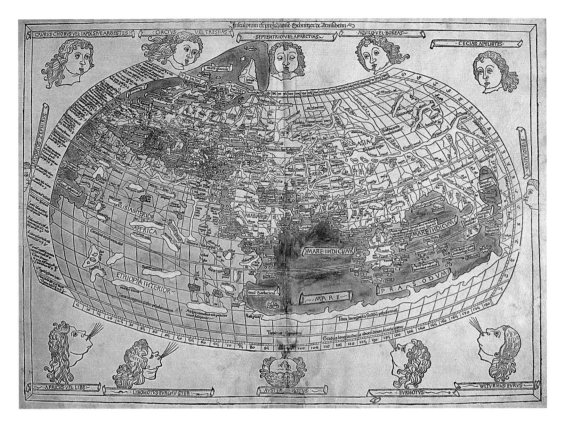

◄ PTOLEMY'S ROUNDED VIEW

This 1482 edition of Ptolemy's second-century AD world map gives a detailed picture of Europe and the Mediterranean and Atlantic coasts of North Africa, with the Black and Caspian seas clearly recognizable. Unknown realms shown encircling the south make the Indian and Atlantic oceans mighty inland seas. This view of the world replaced the flat-earth beliefs of the Dark Ages.

▼ CONTINENTS TAKE SHAPE

Drawn in 1570 by Ortelius, this map records the discoveries of Diaz, da Gama, and later Portuguese navigators of the coasts of Africa, India, East Asia, and Japan; of Columbus and Vespucci of America's east coast; and of Magellan of South America and the Pacific Ocean. *Terra Australis*, seen as a separate continent, includes Australia, New Zealand, and Tierra del Fuego.

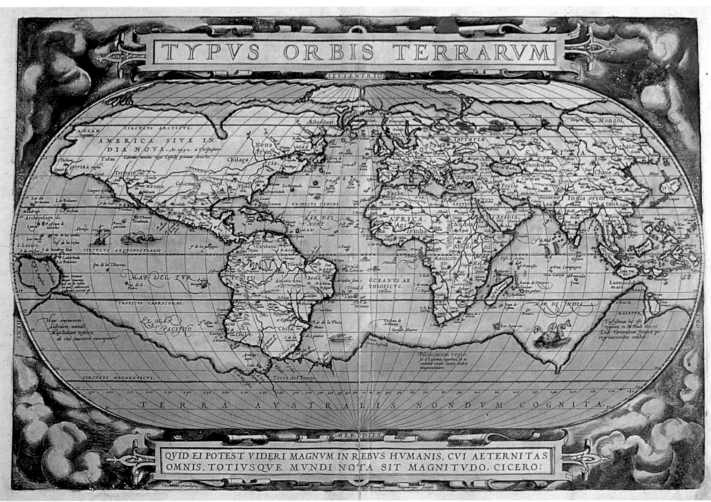

Early navigators

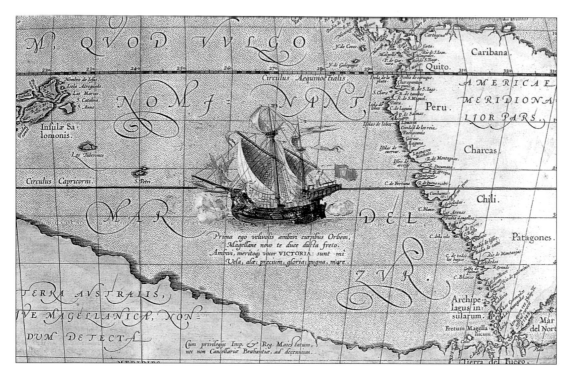

Most early exploration of Antarctica was a process of whittling down the fabled Great South Land as empirical knowledge replaced conjecture. The Portuguese explorer Bartolomeu Diaz took the first significant step when he sailed down the west coast of Africa in 1487. At the time, it was still credible that Ptolemy was right and that *Terra Incognita* filled the bottom of the world with a coast across the temperate zone, so that the Atlantic and Indian oceans were in effect inland seas. It was a brave venture by Diaz, because other ancient predictions stated that humans could not survive the "torrid zone" at the equator. In January 1488 he was alongside the coast of what is now South Africa when storms drove him out to sea and he sailed south for a few days. When he turned east, he found no coast, and made landfall only by sailing north again. He had rounded the bottom of Africa. He realized that he had discovered a sea route to India, and that Africa, at least, was not joined to the Great South Land. When he returned there was talk of his leading a voyage to India, but it was another Portuguese explorer, Vasco da Gama (c.1469–1525), who sailed to India in 1497, thus proving that the Atlantic and Indian oceans were not landlocked. In March 1500 the Portuguese navigator Pedro Cabral led a fleet of 13 vessels to India, but to avoid the becalming waters of the Gulf of Guinea they sailed southwest far enough to see (and claim) the land now known as Brazil.

Next came Amerigo Vespucci. In 1501 he led a Portuguese expedition that reached Brazil in January 1502 and went south to the River Plate. He probably sailed south along Argentina's Patagonian coast. He concluded that he had discovered not a new route to Asia

but a whole new world that was to become known as America. Clearly the world was much larger than Ptolemy had calculated, and this was confirmed when, on 25 September 1513, the Spanish explorer Vasco Núñez de Balboa crossed the Isthmus of Panama and became the first European to see the vast Pacific Ocean.

Ferdinand Magellan

Magellan traveled east as far as Malacca (in what is now Malaysia) with the Portuguese fleet before seeking service elsewhere. He persuaded the king of Spain to send him on a voyage westward to prove that the spice islands —the Moluccas, now part of present-day Indonesia—could be considered part of Spain's western hemisphere of influence. To accomplish this he had to find a passage around the bottom of South America. His five ships left Spain in 1519. From Rio de Janeiro they sailed south, and Magellan explored the River Plate to see if it led through to the Pacific. He continued southward, continually seeking a westward passage and continually being baffled. Finally, he rounded the Cape of the Virgins (Cabo Vírgenes) at 52°50′S and entered the strait that now bears his name.

The Strait of Magellan threads a tortuous and treacherous 525 kilometers (330 miles) between the island of Tierra del Fuego and the South American continent. Magellan sailed through, but could not decide whether Tierra del Fuego was the tip of a continent or an island. It took 38 days to reach the ocean that came to be called the "Pacific" because it seemed so peaceful. The fleet discovered that it was much larger than expected: they had 14 weeks of near-starvation before struggling into their next landfall at Guam. Magellan was killed on 27 April 1521 in the Philippines, but his surviving ships reached the Moluccas (and, in 1522, Spain), thus proving Magellan's thesis that the spice islands could be reached by sailing either east or west, and proving conclusively that the earth was round.

Sir Francis Drake

In 1572, the brilliant English navigator Francis Drake (c.1540–96) led an expedition to raid Spanish possessions in Panama. Here he had his first glimpse of the Pacific. In December 1577 he sailed with five vessels, to map the coast of the South Land that was believed to lie below the Pacific Ocean. He reached the entrance of the Strait of Magellan in August 1578.

When he arrived at the Pacific Ocean in September, Drake did not experience the same benign conditions as Magellan. Rather, an "intolerable tempest" drove the 100-tonne *Golden Hind* southeast as far as 57°S. So, by chance, and traveling backward, Drake discovered the place where the Atlantic and Pacific oceans "meete in most large and free scope" in what is now the Drake Passage. It was clear that Tierra del Fuego was an island, and that the tip of the Southern Continent must lie even further to the south.

In 1579 the Low Countries broke away from Spanish rule. It was the birth of a new seafaring nation, and the Dutch East India Company, founded in 1602, was to have a profound influence on exploration and trade for the next 200 years. By the early seventeenth century it was illegal for ships other than the Company's to pass through the Strait of Magellan, so when the wealthy Amsterdam merchant Isaac le Maire organized a private expedition, he decided to seek out the southern seaway of which Drake spoke.

So in 1615 two vessels, the *Eendracht* and the *Hoorn*, commanded by Wilhelm and Jan Schouten respectively, set sail for Patagonia under the overall command of le Maire's son, Jacob. Fire destroyed the *Hoorn* during the voyage but on 29 January 1616 the *Eendracht* became the first ship to round Cape Horn, which Wilhelm Schouten named Kaap Hoorn after his birthplace in northern Holland. One of the world's most notorious nautical landmarks was now on the shipping maps.

By 1642, the Dutch colony of Batavia was well established and ships of the Dutch East India Company had explored part of the west coast of Australia. But no one knew whether Australia was part of the legendary Great South Land, so Abel Janszoon Tasman sailed from Batavia to explore the Indian and Pacific Oceans. Over 10 months he reached about 49°S and proceeded north to discover Tasmania, which he described as "not cultivated but growing wild by the will of God." He continued to New Zealand, which he assumed was the western shore of the southern continent, and returned to Batavia having circumnavigated the continent of Australia without ever seeing it. The Great South Land had shrunk again.

Over the following century various national expeditions discovered land that they thought was the southern continent but turned out to be small sub-Antarctic islands. James Cook of the Royal Navy undertook the next major exploration of Antarctic waters.

Strange birds

In his passage through the Strait of Magellan, Francis Drake landed on an island, and Francis Fletcher described a "great store of strange birds which could not flie at all," their body size "less than a goose," that laid their eggs on the ground. Fletcher continued: "... in the space of one day, we killed no less than 3000 ... they are a very good and wholesome victuall ..." They were, of course, penguins, though no one on Drake's voyage used that name. It had been applied to the now extinct Great auks of the Arctic—there was a Penguin Island off Newfoundland by 1578— but the name seems to have been first used to refer to the familiar black and white birds of Antarctica by Francis Petty during a circumnavigation of the globe by Englishman Thomas Cavendish in 1586–88; Petty noted that "we trimmed our saved pengwins with salt for victual."

▶ WHERE OCEANS MEET
An "intolerable tempest" drove Drake south into the stormy passage between Tierra del Fuego and the Antarctic Peninsula that now bears his name. Here, viewed from a modern cruise ship, *Akademik Ioffe*, a Force 12 gale whips up surging seas in the narrow gap. How much more overwhelming must such mighty swells have seemed from the tiny *Golden Hind*.

First to cross the Circle

When the Royal Society asked the Admiralty in 1768 to commission a voyage to the Pacific to observe the transit of Venus across the Sun, James Cook was their rather surprising choice of leader. Alexander Dalrymple, who optimistically expected the Great South land to begin south of the tropics and be populated with 50 million inhabitants, had been lobbying to lead a voyage south. Cook, born in Yorkshire on 27 October 1728, was an experienced seaman and a self-taught mathematician and navigator, but was largely unknown. He was also commissioned to find *Terra Australis Incognita*. He left Plymouth in HMS *Endeavour* on 26 August 1768 and returned on 12 July 1771, having observed the transit, claimed New Zealand and the east coast of Australia for Britain, and mapped a great deal of the Pacific Ocean.

Cook's first voyage, successful though it was, did not discover the fabled Southern Continent. Naturalist Joseph Banks deduced from his observations: "That a Southern Continent exists I firmly believe," while Cook concluded that "as to a Southern Continent I do not believe any such thing exists unless in a high latitude." But he was determined to resolve the issue once and for all, and with his second venture a question more than two millennia old would be answered.

The second voyage

In July 1772 Cook, now a commander, set out with HMS *Resolution* and *Adventure* with instructions from the Admiralty for "prosecuting [his] discoveries as near

to the South Pole as possible." First, though, he was to look for Cape Circumcision, mapped by French explorer Jean-Baptiste Bouvet de Lozier in 1739 and potentially the tip of the Southern Continent. The expedition was equipped with provisions for 18 months and the most up-to-date chronometer. Banks had fallen out with Cook and the Admiralty over such matters as accommodation for himself and his party of 17, which included two French horn players; instead, Cook had the erudite but demanding and difficult John Reinhold Forster, described by a Cook biographer as "one of the Admiralty's vast mistakes."

Cook sailed south from Cape Town but could not find Cape Circumcision. Exploring south and east, he realized that it could not have been part of a continent, and wrote dismissively: "I am of the opinion that what M. Bouvet took for land … was nothing but mountains of ice surrounded by field ice." On 17 January 1773 *Resolution* and *Adventure* became the first ships to cross the Antarctic Circle, but Cook ventured only a short distance inside the Circle before encountering heavy ice and retreating to the northeast, little realizing that he was only 130 kilometers (80 miles) from the Antarctic continent. Summer was over so he sailed directly to New Zealand.

Cook's two ships had lost touch in a storm off New Zealand's Cape Palliser, but Cook turned the *Resolution* south in November 1773. On 20 December he crossed the Antarctic Circle at about longitude 148°W. He had seen the first iceberg, but continued eastward through heavy ice. He turned north to explore ocean not covered on his previous voyage, but turned south again at 122°W—to the disappointment of his crew, who thought they were on their way to Cape Horn and home. Instead they crossed the Circle again on 26 January 1774. Over the next few days, they dodged north and south through thick fog and heavy ice. On 30 January

▲ **COOK IN THE ANTARCTIC**
James Cook made the first circumnavigation of Antarctica during the summers of 1772–75, crossing the Antarctic Circle four times. Returning to England he declared "no man will venture further south than I have done, and the lands which may lie to the South will never be explored."

◄ **ISLANDS OF ICE**
The painter William Hodges was employed by Cook to make a visual record of the expedition. His images provided the first glimpse of Antarctica's picturesque but hostile environment. Here, beside a huge, castellated "ice island," men in the ship's boats collect ice for water and shoot seabirds for food, while *Resolution* stands by.

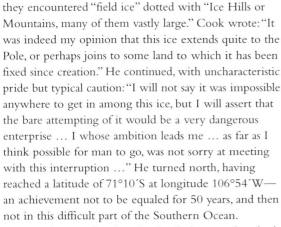

TIMELESS PEAKS, SOUTH GEORGIA

Heading back to England in January 1775, Cook discovered the island of South Georgia, which he named for the king and claimed for Britain. Possession Bay, with rearing jagged mountains behind, is a scene that has changed little since recorded then by William Hodges. However, although King penguins now preen undisturbed on the beach, Cook's reports of the island's abundant sea mammals unleashed a century of exploitation that virtually exterminated fur and elephant seals, and later whales, from these shores and waters.

they encountered "field ice" dotted with "Ice Hills or Mountains, many of them vastly large." Cook wrote: "It was indeed my opinion that this ice extends quite to the Pole, or perhaps joins to some land to which it has been fixed since creation." He continued, with uncharacteristic pride but typical caution: "I will not say it was impossible anywhere to get in among this ice, but I will assert that the bare attempting of it would be a very dangerous enterprise … I whose ambition leads me … as far as I think possible for man to go, was not sorry at meeting with this interruption …" He turned north, having reached a latitude of 71°10′S at longitude 106°54′W— an achievement not to be equaled for 50 years, and then not in this difficult part of the Southern Ocean.

Over the next few days Cook sailed eastward on both sides of the Antarctic Circle before winter (and pack ice) drove him back to warmer regions. Faced with impene-

trable ice, Cook and his crew willingly retreated, weary of the "dangers and hardships, inseparable with the navigation of the Southern Polar regions." He sailed south again from New Zealand in November 1774, heading for the polar spring thaw. For five weeks, *Resolution* sailed east toward Cape Horn, and on 28 December the ship rounded the Horn and continued eastward through the South Atlantic in search of the land reported by London-born merchant, Antoine de la Roche, in 1675. Cook found it—but far from where de la Roche had placed it. It had been considered a potential promontory of the Southern Continent, but at its southwestern point Cook realized it was an island. He named the point Cape Disappointment, and the island South Georgia.

From there Cook followed the 60° latitude through fog but in a sea clear of ice. On 27 January he encountered an iceberg, the harbinger of a lot of sea ice.

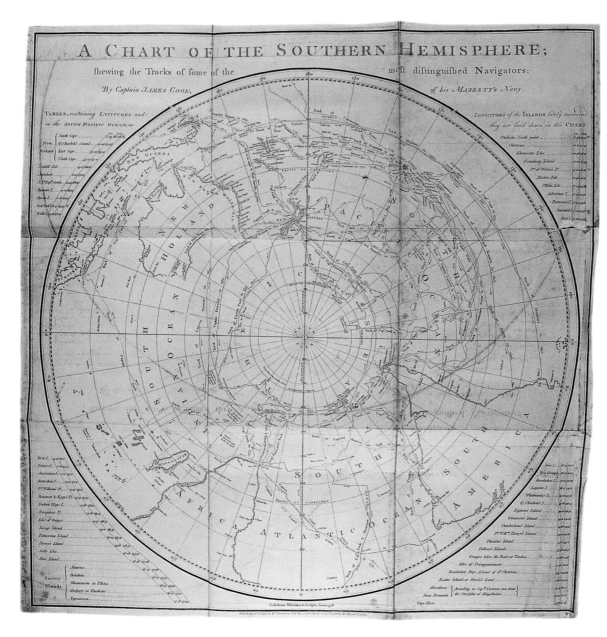

A CHART OF THE SOUTHERN HEMISPHERE;
shewing the Tracks of some of the most distinguished Navigators:
By Captain JAMES COOK, of his MAJESTY's Navy.

◄ COOK'S SECOND VOYAGE
George Forster, son of the
expedition's naturalist, published this map in 1777, only
weeks before Cook's own
"official" map of his discoveries appeared. It shows the
route of the expedition, and
details individual sightings by
Cook in *Resolution* and by his
deputy Furneaux in
Adventure. Cook's second
voyage finally established the
true extent of *Terra Australis
Incognita*—it had to lie south
of the Antarctic Circle.

one of them by sickness."
The first circumnaviga-
tion of Antarctica had
been achieved. The quest
for the mythical bounty
of the fertile great
Southern Continent was
over: it did not exist.
Cook wrote: "I had now
made the circuit of the
Southern Ocean in a high
latitude … in such a man-
ner as to leave not the
least room for the possi-
bility of there being a
continent, unless near the
Pole and out of the reach
of navigation …"

He saw some islands and, tantalisingly, some mountain
peaks, but in the fog and sea ice he could not discern
whether they were located on islands or on the point of
a large landmass. He optimistically named it Sandwich
Land, after the first Lord of the Admiralty. Later he
wrote of this land and ice: "I firmly believe that there is
a tract of land near the Pole, which is the
source of most of the ice which is spread
over this vast Southern Ocean … I can be
bold to say, that no man will ever venture
farther than I have done and that the lands
which may lie to the South will never be
explored. Thick fogs, snow storms, intense
cold and every other thing that can ren-
der navigation dangerous one has to
encounter and these difficulties are greatly heightened by the inexpressible horrid aspect of the country, a country doomed by nature never once to feel
the warmth of the sun's rays, but to lie forever
buried under everlasting snow and ice."

Cook returned to England on 30 July 1775,
three years and eighteen days after his departure, "In
which time," he noted, "I lost but four men and only

The third voyage

Cook's third Pacific voyage once more took him into
Antarctic waters. He sailed on the *Resolution* again, this
time accompanied by the *Discovery*. The stated purpose
was to return a native of Tahiti to his home, but the
scope of the voyage was much wider: he was to find the
Northwest Passage—the long-sought
northern shortcut from Europe to
Asia. He sailed via Cape Town to
inspect some islands discovered by
French captain Yves-Joseph de
Kerguelen-Trémarec in 1772 in the
southern Indian Ocean. Today they are
called Iles Kerguelen, but Cook
thought Desolation Islands a more
apt name for them. He proceeded
through the Pacific, along the west
coast of America and through the
Bering Strait to cross the Arctic
Circle. On the voyage north he had
discovered the Hawaiian Islands
(which he named the Sandwich
Islands), and when the Arctic winter

set in he returned there. On 14 February 1779 he intervened when one of his boats was stolen. A fight ensued, and Cook was killed. It was indeed a tragic end to a remarkable life.

By happy chance, William Wales, the astronomer on *Resolution*, went on to teach at the Mathematical School at Christ's Hospital in London. Young Samuel Taylor Coleridge was one of his pupils in the 1780s, and the schoolmaster's account of ice and albatrosses inspired the poet's later masterpiece, *The Rime of the Ancient Mariner*, which vividly and remarkably accurately evokes an environment he experienced only through hearsay.

Ne plus ultra

The contesting boasts of two men who were on the *Resolution* when she reached Cook's southernmost latitude on 30 January 1774 must have entertained the crew on the long voyage home. English explorer George Vancouver waited until the ship was ready to tack about, and then climbed to the end of the bowsprit to exclaim *Ne plus ultra* ["No further is possible."] But Swedish doctor and naturalist Anders Sparrman had a rival claim: "I went below [to his stern cabin] ... to watch ... the boundless expanses of Polar ice. Thus ... I went a trifle farther south than any of the others ... because a ship ... always has a little stern way before she can make way on a fresh tack."

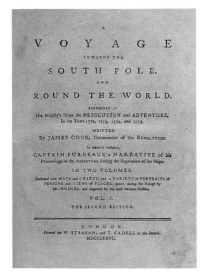

▲ MAGNIFICENT FORMS

An engraving from *The Three Voyages of Captain Cook Around the World*, published in about 1779. Cook's small wooden *ships— Resolution* and *Adventure* were only about 30 meters (100 ft) long—would have been dwarfed by the huge islands of ice they encountered in Antarctic waters.

▲ MOMENTOUS WORK

The narrative of Cook's second voyage, published in 1777, was evocatively but accurately subtitled "Towards the South Pole and around the World." It was an epic that established the framework for virtually all subsequent Antarctic exploration.

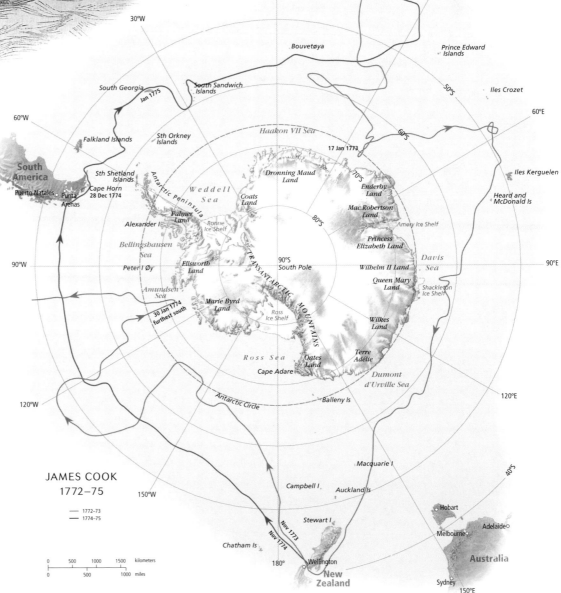

JAMES COOK
1772–75

—— 1772–73
—— 1774–75

0   500   1000   1500   kilometers
0   500   1000   miles

The quest for the South Pole

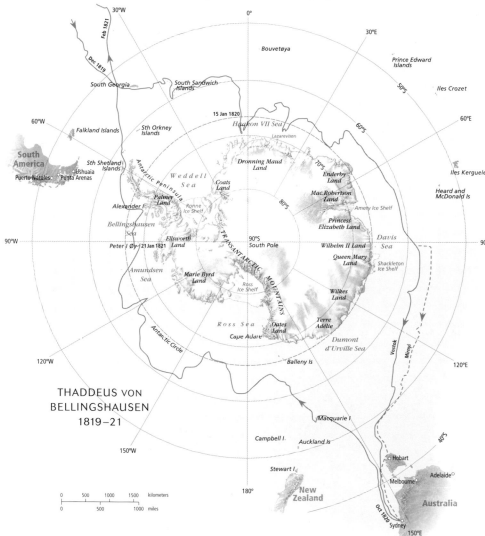

Cook died in 1779, and the world had to wait 40 years for another explorer to match his achievements. But in May 1819 the Russian explorer Thaddeus Thaddevitch von Bellingshausen took command of an expedition to Antarctica. Hugh Robert Mill, historian of the Heroic Age of Antarctic exploration, described this as "one of the greatest Antarctic expeditions … well worthy of being placed beside that of Cook … a masterly continuation of Cook, supplementing it in every particular, competing with it in none."

Born in 1778 and a naval cadet from the age of 10, Bellingshausen had served as fifth lieutenant on the first Russian voyage around the world, led by Admiral Adam Johann, Baron von Krusenstern from 1803 to 1806. When the Russian Antarctic expedition was being planned as part of a burgeoning nationalism, Bellingshausen was chosen to command the two ships, *Vostok* and the smaller and slower *Mirnyi*, with a total of 189 officers and crew. He was allowed very little time to prepare for this monumental undertaking—indeed, the expedition sailed in late July 1819, about two months after he was recalled to St Petersburg from survey work in the Black Sea. His instructions were succinct: to build on the explorations of Cook, whom the Russians greatly admired, and "to approach as closely as possible to the South Pole, searching for as yet unknown land, and only abandoning the undertaking in the face of insurmountable obstacles." The quest for the South Pole (rather than the Great Southern Continent) had begun.

A modest hero

Bellingshausen's expedition has often been overlooked in Antarctic history because his logs are lost and his unassuming personal record was available only in Russian until 1902, when a German translation appeared; there was no English version until 1945. Few visitors realize the historical significance of the name of the stretch of the Southeast Pacific to the west of the Antarctic Peninsula: the Bellingshausen Sea.

The two ships left Rio de Janeiro on 20 November 1819 for South Georgia, where Bellingshausen completed Cook's survey by mapping the southern coast. He then followed Cook's route to the South Sandwich Islands and, in better conditions than Cook had experienced, found that they were just more small islands, and that there were more of them than Cook had observed. Continuing eastward, he crossed the Antarctic Circle on 15 January 1820—his was the second expedition to do so—but did not even note this first crossing in his narrative. But while Cook sailed 24° of longitude within the Circle, Bellingshausen was to cover more than 42°.

▲ ICY FRINGE SIGHTED
Bellingshausen came much closer than Cook to the Continent. He sighted what was probably the Lazarevisen in February 1820 and found Peter I Øy in January 1821.

THADDEUS VON
BELLINGSHAUSEN
1819–21

▶ GRAND HARBOR
Deception Island in the South Shetlands group is a drowned volcanic caldera and one of the finest natural harbors in Antarctica. When Bellingshausen sailed through its narrow entrance (on the far left of this photograph) in January 1821, he found 18 British and American sealing ships at anchor, even though the island had only been discovered the previous year.

On 16 January he would have seen the continent if the weather had been fine, but he was sailing through snow when he observed "a solid stretch of ice running from east through south to west." Probably he was viewing an ice shelf at the base of Haakon VII Sea, which may at that time have extended far out to sea. Geographers regard continental ice as part of the land-mass—otherwise, the Antarctic Peninsula would be classified as an archipelago—so the first sighting of "Antarctica" may be credited to Bellingshausen. Some have contended that what he saw was pack ice, but his use of the Russian term for "continental ice" works against that argument.

Land in sight!

Bellingshausen's information for 5 February is more definite. On a day with good visibility he wrote: "The ice to the SSW is attached to cliff-like, firmly standing ice: its edges were perpendicular and formed bays, and the surface rose in a slope towards the south, over a distance whose limits we could not see from the cross-trees." He must have been looking at the Lazarevisen (Lazarev Ice Shelf). But he needed supplies, so he sailed

north to Port Jackson (now Sydney). On 31 October he sailed south again to complete his circumnavi-gation of the South Pole. However, like Cook before him and Shackleton to come, he encountered unseasonably heavy ice as he approached the conti-nent. Thus he missed the Ross Sea and could not cross the Antarctic Circle until 14 December. But on 21 January 1821 he reached the southernmost point of his voyage—69°59′S—and dis-covered Peter I Øy, the first land ever seen within the Antarctic Circle. He also discovered and named Alexander I Land (now Alexander Island), stating: "I call this discovery 'land' because its southern extent disap-peared beyond the range of our vision." It is in fact a large island, separated from the continent by a narrow channel but linked to it by ice. He sailed on to the South Shetland Islands, which he mapped, and on Teille Island (now Deception Island) he encountered American

▲ FROZEN FASTNESS
Peter I Øy, named by Bellingshausen for Russia's greatest czar, is one of the world's most inaccessible islands. The first landing was more than 100 years after discovery, and barely a dozen expeditions have since broached these forbidding shores. Only one has reached the interior—by helicopter.

▲ BELLINGSHAUSEN'S SHIPS
Although here they appear
identical, *Mirnyi* was not only
a much smaller ship, but was
also much slower, which
frustrated Bellingshausen in
Vostok. The creator of this
image had probably not been
south himself—unlike William
Hodges, who painted scenes
of Cook's voyages from
personal experience.

➤ ISLAND PROFILES
Bellingshausen's navigator drew
these careful sketches of the
islands that make up Cook's
"Sandwich Land." They can be
recognized from the same point
of view today. Cook and Thule
Islands are remnants of the rim
of a large volcanic crater, and the
caldera between them, though
open, is of similar formation to
Deception Island.

Bristol Island, 7 miles distant

Thule Island

Cook Island, 15 miles distant

▼ STANDING THEIR GROUND
Bellingshausen called the
South Sandwich Chinstrap
penguins "ringed" or
"common" penguins, and
found that they pursued his
officers aggressively.

and British sealing ships working the island group. On
25 January a young American captain, Nathaniel Palmer,
came aboard; Palmer was later to lay dubious claim to
have been the first to see the Antarctic Continent.

One of Bellingshausen's ships, *Vostok*, had been shipping water since leaving Port Jackson, and as winter
approached Bellingshausen decided to turn north to
Rio de Janeiro at the end of January, arriving
there on 27 March. He overhauled the ships
over the next month and sailed back to
Kronstadt via Lisbon, completing the
voyage on 4 August 1821. The end of
his narrative is typically brief and
factual: "We had been absent for
751 days. During that time we
had been at anchor in different
places 224 days and had been
under sail 527 days. Altogether

we had covered 57,073½ miles … During the course
of our voyage we had discovered twenty-nine islands:
two of these were in the Antarctic, eight in the South
Temperate Zone, and nineteen in the Tropics."

He did not mention that only three men had died
throughout the voyage—a record that Cook would have
admired. And he had achieved the remarkable feat of
circumnavigating Antarctica closer to the coast than
Cook—so close that he was the first to see Emperor
penguins, the largest and southernmost-dwelling of all
penguins. But the voyage had suggested few commercial
possibilities and was largely ignored in Russia; and
because his charts and logs were not accessible to non-
Russian speakers the voyage was overlooked by other
nations. Bellingshausen rose to the rank of admiral
during the next 30 years of his naval career and then
was appointed governor of Kronstadt, the role he still
held when he died in 1852.

The first sight and the first step

The South Shetland Islands, off the west coast of the Antarctic Peninsula, are central to the question of who was the first to "see Antarctica." The islands were discovered in February 1819 by sealer William Smith, whose report of his find triggered the islands' sealing boom. Smith was soon employed by the British Admiralty to survey the islands under the command of Edward Bransfield the following summer.

The first sight

On 30 January 1820 Bransfield and Smith saw and charted part of the Antarctic Peninsula, and named it Trinity Land (now Trinity Peninsula). Exactly two weeks earlier, Thaddeus von Bellingshausen had probably seen the icy fringe of Antarctica, far to the west. However, he had glimpsed ice and they were looking at rocky mountains. In November 1820 Nathaniel Palmer, captain of the sealer *Hero*, sailed through the narrow entrance of Neptune's Bellows and into the spectacular caldera known as Port Foster. He was perhaps the first to do so; Bransfield and Smith had seen Deception Island on 29 January 1820, but did not investigate in the thick fog. Palmer met Bellingshausen in January 1821, and was later quoted as saying: "I informed [Bellingshausen] of … the discovery of land … and it was him that named it Palmers Land," but his only likely sighting is dated November 1820—10 months after Bransfield and Smith.

The first step

The honor of the first step onto the Antarctic Continent probably belongs to the crew of a boat from the *Cecilia*, an American vessel skippered by John Davis. On 7 February 1821 he recorded "… open cloudy weather and light winds a standing for a large body of land in that direction SE at 10 am close in with our boat and sent her on shore … I think this southern land to be a continent." The landing probably took place at Hughes Bay (64°13′S 61°20′W), and much pre-dates the other documented claim, that of the Norwegian businessman Henryk Johann Bull, who led a whaling expedition to the Ross Sea region in 1895. The area is now the Davis Coast, but the names of the crew who made the historic one-hour landing are unknown.

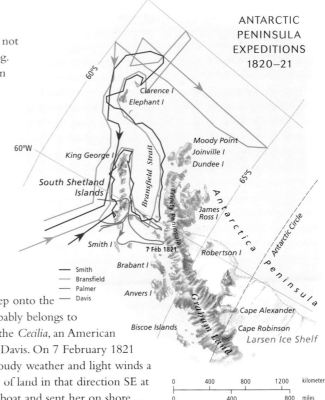

ANTARCTIC PENINSULA EXPEDITIONS 1820–21

— Smith
— Bransfield
— Palmer
— Davis

▼ ANTARCTIC PENINSULA FROM DECEPTION ISLAND
From Deception Island (foreground), the mountains of the Antarctic Peninsula can be seen across Bransfield Strait. British mariners Smith and Bransfield charted part of the Peninsula in January 1820, naming it Trinity Land. Ten months later American sealer Nathaniel Palmer explored Deception Island, and from there spied these icy peaks. He believed his was the first sighting of the Antarctic mainland.

Further south than any before

Travelers to Antarctica still speak of James Weddell with awe. The Weddell Sea—one of the two great indentations into the Antarctic Continent—is a spawning ground of polar ice, yet in 1823 Weddell sailed further south than was conceivable at the time. His record stood for 18 years, until James Clark Ross ventured into the waters on the other side of Antarctica, and steam had ousted sail before anyone replicated Weddell's achievement in the same location.

Weddell was British, but was born in Ostend in 1787. He joined the Royal Navy at the age of nine, and then alternated between the Royal Navy and the Merchant Navy before taking command of the 160-tonne sealer *Jane* in 1819. He returned in 1821 with enough seal skins to buy the even smaller 65-tonne *Beaufoy* to partner *Jane*.

An inauspicious start

The voyage that set Weddell's mark on Antarctic history sailed down the Thames on Friday, 13 September 1822. Directly after the inauspicious date, while still in English waters there was a collision and *Jane* was damaged. The vessels sailed to the South Orkneys but found few seals—and they were unlike any that Weddell had seen before. (He took the skins and skulls back to Robert Jameson of Edinburgh University, who declared a new species: the Weddell seal, *Leptonychotes weddelli*.)

But the immediate task was to find more seals, or more land where they might live. Weddell offered a reward of 10 pounds—most of an able seaman's annual wage—for the first sighting of land. By February 1823 he had concluded that such discovery would only be made to the south. He pushed south into a prevailing wind through intense cold and a sea jagged with ice. It was a miserable voyage, but Weddell's judgment was partially vindicated on 16 February by a change in the weather: the wind shifted to the west, the ice disappeared, and flocks of seabirds were reflected in the calm sea. The next day they reached 71°34′S 30°12′W—no ship had ever been so far south.

A stroke of luck

In these ideal conditions Weddell pressed southward. On 19 February he wrote a paragraph that those who know the area still read with wonder: "In the evening we had many whales about the ship, and the sea was literally covered with birds of the blue peterel kind. NOT A PARTICLE OF ICE OF ANY DESCRIPTION WAS TO BE SEEN … had it not been for the reflection that probably we should have obstacles to contend with in our passage northward, through the ice, our situation might have been envied."

On 20 February 1823 a rising south wind forced Weddell to make a decision. At noon the ship was at 74°15′S 34°16′W and only three icebergs were visible, but winter and the polar night were closing in, and there was ice to be crossed in sailing north. Caution won the day, and Weddell fired the canon, raised the colors, and issued an extra allowance of rum. Weddell named his discovery George IV Sea, but in 1900 it was renamed the Weddell Sea.

Weddell was fortunate in his easy passage into the Weddell Sea—many who followed did not fare so well, even in larger, purpose-built ships. His feat was so difficult to replicate that skeptics doubted his veracity, and he had three of his crew swear under oath that the ship's logs were correct. But James Clark Ross, the next

◄ SKILL AND DARING
Taking advantage of uncharac-
teristically ice-free seas,
Weddell's flagship, the brig
Jane, followed by her smaller
companion the cutter *Beaufoy*,
reached their furthest point
south on 17 February 1823.
Although Weddell made the
most of his opportunities, the
voyage failed in its primary
objective of finding new
sealing grounds; nor did
he discover new lands.

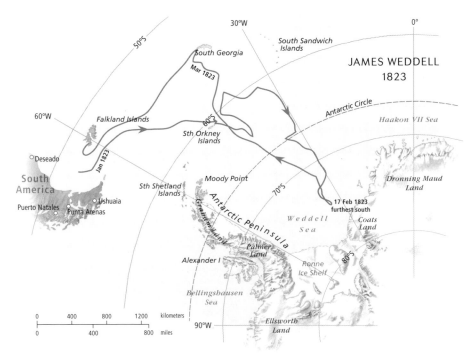

▲ WEDDELL SEAL AT REST

On the South Orkneys Weddell discovered a new species of seal, which was named *Leptonychotes weddelli*. The encounter proved more significant to science than to commerce. Compared with the valuable fur seals found further north, Weddell seals have much poorer, thinner coats, of little economic worth.

▲ AS SEEN BY WEDDELL

Weddell's drawing of a "Sea Leopard" clearly shows the mottled fur, the twinned rear flippers, and horizontal posture of a true seal. Weddell would have been more familiar with the more upright posture and separated rear flippers of the fur seal.

to hold the title of "furthest south" wrote generously that Weddell "was favored by an unusually fine season, and we may rejoice that there was a brave man and daring seaman on the spot to profit by the opportunity."

Jane and *Beaufoy* returned to England in July 1824. In *A Voyage Towards The South Pole*, published in 1825, Weddell wrote: "I have only done that which every man would endeavor to accomplish, who in the pursuit of wealth, is at the same time zealous enough in the cause of science to lose no opportunity of collecting information for the benefit of mankind." *Jane* had to be scrapped in 1829. It was a financial blow, and Weddell died in poverty in London in 1834, aged 47.

Sealers and whalers

It has been said that few did more to endanger Antarctic wildlife than James Cook who told of islands teeming with seals and the Southern Ocean filled with whales. But earlier explorers, including Francis Drake, had also reported this profusion.

In the late seventeenth century the piratical English navigator William Dampier called at the Juan Fernandez Islands, off the coast of Chile, and wrote: "seals swarm around … ." Thus notified, entrepeneurial sealers began to look south. In 1797 throngs of sealers arrived, and by 1807 few seals remained. One vessel took off 100,000 skins, and the total harvest was estimated at three million. Forty thousand skins were shipped to London from the Falkland Islands in 1788, the year the first British sealers reached South Georgia. In 1800, a New York sealer took 57,000 skins from the island, and in 1825 James Weddell calculated that 1.2 million fur seals had been killed on South Georgia. More than 120,000 skins had also been plundered from Macquarie Island.

Biscoe and the Enderby Brothers

As their quarry approached extinction, the sealers sought new hunting grounds—but important discoveries were made during those voyages. John Biscoe, at 36, was an ex-Royal Navy seaman employed by the Enderby Brothers. The Enderby tradition combined exploration with exploitation, and captains were recruited to collect flora and fauna. The policy bankrupted the company, but it left an enduring legacy of Antarctic exploration. Biscoe was to seek out new sealing grounds, but also to make discoveries in high southern latitudes. His vessels were the *Tula* and the much smaller *Lively*, sparsely equipped and provisioned, with 29 crew. They sailed from England in July 1830 and crossed the Antarctic Circle in January 1831. On 28 January they arrived at their furthest south: 69°S at 10°43′E.

Biscoe ventured further south than Bellingshausen, and in a way had better luck. He first definitely sighted land on 28 February 1831 at 66°S 47°20′E, when

▼ RETURN FROM THE BRINK
This fur seal pup at Cooper Bay, South Georgia, represents the revival of fur seals after their virtual extinction in the early nineteenth century. Weddell calculated that by 1825, 1.2 million fur seals had been slaughtered on the beaches of South Georgia.

"several hummocks" resolved themselves into "the black tops of mountains … through the snow." But after two days of trying to land, Biscoe gave up. He named his discovery Enderby Land. On 3 March his ships were separated by a three-day storm that drove *Tula* north. The indefatigable Biscoe turned his damaged vessel south again, hoping that the ice had blown away from the shore. It had not, and with some crew injured and others sick with scurvy, he sailed north. When *Tula* reached Hobart, only Biscoe and four others could stand, and two of the crew had died.

Despite harrowing experiences and the deaths of seven more crew, Biscoe's expedition continued for many months. In late February 1832 they arrived in the South Shetland Islands, having completed a circumnavigation of Antarctica and finally established that there was a large continent at the heart of the ice. Biscoe wrote: "I am firmly of the opinion that this is a large continent as I saw to an extent of 300 miles."

Biscoe needed to find seals before winter set in, but he failed, and *Tula's* rudder was damaged in a late storm. He retreated to the Falklands, where *Lively* was wrecked. He wanted another season in Patagonia to make the voyage a financial success, but one by one his hard-pressed crew deserted. The bedraggled *Tula* sailed into London on 8 February 1833 with just 30 seal skins to show for a 30-month voyage.

Enduring legacy

Little is known about British sealer and whaler John Balleny, but his name lives on in a group of five islands rising sheer from the water and lying across the Antarctic Circle below New Zealand. The *Eliza Scott*, under Balleny, and the much smaller *Sabrina*, under Thomas Freeman, sailed from England in July 1838 via the Cape of Good Hope and Iles Amsterdam to New Zealand. On 1 February 1839 they reached 69°02′S at 172°11′E, but the way south was blocked by ice. However, it showed that there was a possible route along this longitude: the way into the Ross Sea had opened a crack. On 9 February Balleny saw what are now the

SEALERS AND WHALERS 1831–39

islands that bear his name, and he briefly landed there on 12 February; it was the first landing below the Antarctic Circle.

While early sealing brought wildlife populations to the verge of extinction, early whaling was less successful. The whaling industry's chance for mass destruction of species other than Southern right whales would have to wait for the invention of the explosive harpoon head and the factory ship. However, those few early whalers and sealers who pursued new lands with as much enthusiasm as they hunted their quarry left an enduring legacy in Antarctic exploration.

A Commercial venture

△ CAPTAIN OF COMMERCE
After several voyages to the Weddell Sea, Norwegian Carl Anton Larsen (1860–1924) went on to found South Georgia's lucrative whaling industry. Using explosive harpoons, Larsen became one of the most successful whaling captains of the early 1900s.

The powerful Australian interest in the Antarctic Continent can be traced back to Henryk Johan Bull, a Norwegian who had emigrated to Melbourne in 1885. In *The Cruise of the Antarctic*, Bull reported his disappointment when an attempt by Australian scientists to raise an Antarctic expedition failed, and his belief that "an expedition on commercial lines would possibly find supporters."

Whaling in Antarctic waters went back to the early nineteenth century, but at that time there were enough whales in the northern hemisphere to discourage the longer and more hazardous voyage south. Northern whale populations began to fall and alternative products such as mineral oils emerged, but the demand for baleen from right whales—mainly for the "whalebone" used in corsets—kept the quest alive. In 1873 a German expedition explored the Antarctic Peninsula but found only rorqual whales. Over the summer of 1892–93 a Scottish expedition explored the northwest Weddell Sea but found no right whales among the many it saw. A Norwegian expedition under Carl Anton Larsen worked the same area that summer, but with little success in terms of whaling.

Larsen returned the next summer and had more luck with seals than whales, although his voyage was notable because he found petrified wood at Cape Seymour— the first fossils dicovered on Antarctica.

▷ HAZARDS OF ICE AND STORM
HAZARDS OF ICE AND STORM
The *Antarctic* dodging ice on its way to Cape Adare where Henryk Bull—with several others—made the first landing on East Antarctica. Bull was unsuccessful, however, in his primary task of locating baleen whales. On a later expedition in the Weddell Sea *Antarctic* was crushed by ice and sank.

Raising the money

Larsen named Foyn Land (now the Foyn Coast) in 1893 after Svend Foyn, the Norwegian inventor of the explosive harpoon, who went on to mount the new device on the steam-driven catchers that had replaced rowing boats. These innovations revolutionized the whaling industry and made Foyn rich. In 1893, when Bull failed to raise finance in Melbourne, he went to Norway to see Foyn who promised him a ship—a 226-tonne steam whaler named *Kap Nor*, renamed *Antarctic* for the voyage. The ship ran aground on a preliminary whaling expedition to New Zealand's Campbell Island, but sailed from Melbourne Wharf on 26 September 1894, called at Hobart, and set off for Antarctica on 13 October. Her captain was Leonard Kristensen: Bull was on board in the role of manager.

The expedition posed little threat to Antarctica's whale population, but it added to the controversy about who first set foot on the Antarctic Continent—if

CARL LARSEN
1893–94

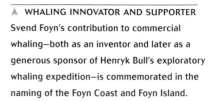

△ WHALING INNOVATOR AND SUPPORTER
Svend Foyn's contribution to commercial whaling—both as an inventor and later as a generous sponsor of Henryk Bull's exploratory whaling expedition—is commemorated in the naming of the Foyn Coast and Foyn Island.

John Davis had not in fact landed on the Antarctic Peninsula in 1821, then Bull's party was the first recorded landing on the continent. The date was 24 January 1895; Bull wrote: "Cape Adare was made at midnight. The weather was now favorable for a landing, and at 1 am a party, including the Captain, second mate, Mr. Borchgrevink, and the writer, set off, landing on a pebbly beach of easy access after an hour's rowing through loose ice, negotiated without difficulty … The sensation of being the first men who had set foot on the real Antarctic mainland was both strange and pleasurable, although Mr. Foyn would no doubt have preferred to exchange this … for a right whale …"

Paving the way for the Heroic Age

The *Antarctic* expedition was an important development in the exploration of Antarctica because the voyage opened the way for the land explorations of the Heroic Age. In *The Cruise of the Antarctic*, Bull wrote: "We have proved that landing on Antarctica proper is not so difficult as it was hitherto considered, and that a wintering party have every chance of spending a safe and pleasant twelvemonth at Cape Adare, with a fair chance of penetrating to, or nearly to, the magnetic pole by the aid of sledges and Norwegian skis."

The immediate commercial returns from the voyage were few: it had found no baleen whales, and while *Antarctic* had been exploring its namesake, its owner, Svend Foyn, had died, so that soon after it returned to Melbourne on 12 March 1895 the ship was instructed to return to Norway. But the expedition is justly remembered in scientific circles, and there is now a Bull Island in the Possession Islands group. This was where the crew of *Antarctic* landed and found lichen—the first vegetation ever discovered in Antarctica, an environment previously believed to be too bleak to allow any plant life to survive.

> **EXPERIENCED HAND**
Carsten Borchgrevink, one of the claimants to the honor of being first to "set foot in South Victoria Land," had been employed by Bull as a seal shooter and "useful hand." In 1901 he returned to Antarctica with his own expedition ship, *Southern Cross* (shown here).

A scientific adventure

▲ DE FACTO LEADER
When *Belgica* became ice-bound, ship's doctor Frederick Cook devised a regime of diet and activities to safeguard the physical and mental health of the crew. He also led the sawing of a channel through the ice to open water and freedom.

At the Sixth International Geographical Congress in London, July 1895, Clements Markham said: "The exploration of the Antarctic regions is the greatest piece of geographical exploration yet to be undertaken."

Meanwhile the Brussels Geographical Society was assembling an Antarctic scientific expedition initiated and led by a young naval lieutenant, Adrien Victor Joseph, Baron de Gerlache de Gomery. The venture had an international flavor. In Norway, de Gerlache bought *Patric*, a 250-tonne whaler, and renamed it *Belgica*. His crew included a Rumanian chief scientist, Emile Racovitza, Henryk Arçtowski, a Pole, as geologist, and a 25-year-old Norwegian, Roald Amundsen, as first mate. Ship's doctor was American Frederick Cook.

Belgica left Antwerp on 16 August 1897 with new cabins and a new laboratory. The plan was to sail along the eastern side of the Antarctic Peninsula, to winter in Melbourne, and to visit Victoria Land the following summer. They reached Punta Arenas early in December, but de Gerlache decided to explore Tierra del Fuego, so they didn't arrive off the Antarctic Peninsula until 20 January —almost the end of summer.

Ice-bound

Over the next few weeks they explored the islands that line what is now Gerlache Strait. They took some of the first photographs of Antarctica, and Cook recorded one occasion as "perfectly dazzling … a photographic day." They crossed the Antarctic Circle on 15 February and met pack ice, but pushed on until it became clear they were trapped in ice for the winter. Amundsen wrote that they "faced the prospect of a winter in the Antarctic with no winter clothing … without

adequate provisions … and even without lamps enough … It was a truly dreadful prospect."

Pack ice can move as fast as 16 kilometers (10 miles) a day, and it carried *Belgica* westward. De Gerlache had sailed well southwest of the Antarctic Peninsula, and *Belgica* became trapped at 71°30′S 85°16′W on 2 March 1898; she was not freed until 14 March 1899, at 70°30′S 103°W. It was a harrowing time. The sun sank below the horizon on 17 May, not to reappear for 70 days.

Cold storage

Cook described the momentous occasion when the bright top of the sun reappeared: "For several minutes my companions did not speak … we could not … have found words with which to express the buoyant feeling of relief, and the emotion of the new life which was sent coursing through our arteries." His optimism was premature: summer progressed, but no way out of the ice presented itself and they faced the prospect of another winter trapped in the ice with little hope of survival. Antarctica's history contains some of the most harrowing tales of adventure imaginable, and the *Belgica*'s began on New Year's Eve when they saw open water 640 meters (2,100 ft) away. Cook suggested cutting a passage to freedom, although they had only three four-foot saws. It was a desperate measure, but they worked day and night (indiscernible during this season of perpetual sunlight) for a month, cutting back toward the ship from the basin. By the end of January they had hacked through over 600 meters (2,000 ft) of ice. Then, as Amundsen wrote: "Imagine our horror on awakening to discover that the pressure from the surrounding ice pack had driven the banks of our channel together, and we were locked in as fast as ever."

Escape

Not surprisingly, depression set in—but that, too, was to prove unjustified. On 15 February the ice relented and the passage they had cut reopened and extended right to the ship. They took *Belgica* to the basin under power, but they were still 11 kilometers (7 miles) from freedom and winter was approaching. Amundsen wrote: "Other weary weeks passed … Then the miracle happened—exactly what Cook had predicted. The ice opened and the lane to the sea ran directly through our basin! Joy restored our energy, and with all speed we made our way to the open sea and safety." It was 14 March 1899, and *Belgica* returned to Antwerp on 5 November 1899.

Despite its problems, the expedition gathered much valuable scientific data. Arçtowski believed correctly that his bathymetric measurements suggested the Antarctic Peninsula was an extension of the Andes—not directly across the Drake Passage but through the Scotia Arc. And, although the feat was achieved unwillingly, this was the first expedition to collect meteorological readings below the Antarctic Circle through a winter. **DM**

Map labels:

60°W

Falkland Islands

30°W

Sth Orkney Islands

60°S

Antarctic Circle

South America

Comodoro Rivadavia
Deseado

Tierra del Fuego
Sth Shetland Islands

Moody Point

70°S

Puerto Aisén
Puerto Natales
Ushuaia
Punta Arenas

Gerlache Strait

Antarctic Peninsula

Weddell Sea

Coats Land

Alexander I

Palmer Land

Ronne Ice Shelf

80°S

ADRIEN DE GERLACHE 1898–99

trapped in ice 2 Mar 1898

Bellingshausen Sea

90°W

Ellsworth Land

Peter I Øy

freed 14 Mar 1899

Amundsen Sea

Marie Byrd Land

120°W

| 0 | 400 | 800 | 1200 | kilometers |

| 0 | 400 | 800 | miles |

FROZEN MEMORIAL

Arriving on the west coast of the Antarctic Peninsula in late summer, the expedition explored and mapped the islands and coast bordering what is now Gerlache Strait (above). The mountainous coast commemorates Emile Danco, the expedition's geophysicist, who died later on the voyage, and Wiencke Island is named after a young seaman who fell overboard and drowned.

TRAPPED FOR THE WINTER

As the long, dark winter deepened, so did the snow drifts around *Belgica*. Some pressure ridges developed, but the ship was never threatened. On moonlit nights the crew went on skiing excursions to relieve the cramped monotony of life on board, while the ship, its rigging glistening with ice, gleamed like a ghost.

The Heroic Age

1901–17

1901–04: Toralf Grunden, Gunnar Andersson, and Samuel Duse (left to right), trapped at Hope Bay for nine months, supplemented their meager supplies with any penguin, seal, or fish they could catch and cook. Their unshaven faces are blackened by blubber and soot.

1901–04: The *Morning* was the relief ship for the *Discovery* expedition. Crewman Doorly wrote: "There are few oceans so tempestuous as that globe-encircling expanse to the south-ward ... usually known as the Southern Ocean."

1901–04: Otto Nordenskjöld (1869–1928) was a university lecturer in geology and mineralogy at Sweden's Uppsala University. Before going to Antarctica, he had already led a successful expedition to research glacial geology in Patagonia.

1901–04: On 23 June 1902 Scott wrote: "The mess-deck was gaily decorated with designs in colored papers and festooned with chains and ropes of the same material, the tables loaded with plum pudding, mince pies, and cakes ... we left the men to enjoy their Christmas fare with an extra tot of grog."

1901–03: Drygalski took this photograph of *Gauss* under full sail. The ship was purpose-built and the specifications were sent to six shipyards for tender. The *Gauss* was a three-masted barquentine (with an auxiliary engine) of similar design to the *Fram*.

1901–03: Erich Dagobert von Drygalski (1865–1949) was born in Prussia in what is now Kaliningrad. His interest in glaciology began in Greenland and took him to Antarctica and later to Spitsbergen. He spent many years meticulously preparing his expedition research for publication.

Erich von Drygalski leads the first German Antarctic Expedition in the ship *Gauss*. They explore East Antarctica and in February 1902 discover and name an area of high ice cliffs for Kaiser Wilhelm II.

On 4 February, Robert Scott becomes the first person to fly over Antarctica when he rises in a tethered hydrogen balloon, at Balloon Bight (now the Bay of Whales).

In the *Scotia*, William Spiers Bruce and the members of the Scottish National Expedition sight the coast of Coats Land. After further exploration, Bruce deduces that this long coastline must be an extension of Enderby Land, and therefore part of the continent rather than just an island—a significant discovery.

The British Antarctic Expeditioners, under the leadership of Ernest Shackleton, set up a base at Cape Royds on Ross Island from the *Nimrod*. Six members of that wintering party achieve the first ascent of Mount Erebus on 10 March. By 26 November, Shackleton again claims a record for the furthest south.

Charcot leads another scientific expedition to Antarctica, this time in the *Pourquoi-Pas?*. The expedition charts 2,000 kilometers (1,240 miles) of coastline and gives form to the whole western side of the Antarctic Peninsula, as well as collecting so much scientific data that it takes a decade and 28 volumes for it to be published.

| 1901–04 | 1901–03 | 1901–04 | 1902 | 1902 | 1902–04 | 1903–05 | 1908 | 1909 | 1908–10 |

Important scientific discoveries are made on the British National Antarctic Expedition (or "*Discovery* expedition") to the Ross Sea, led by Robert Falcon Scott in the ship *Discovery*.

The Swedish South Polar Expeditioners, led by Nils Nordenskjöld, live through two Antarctic winters. They carry out scientific studies and map the eastern side of the Antarctic Peninsula.

The first ever attempt to sledge to the South Pole leaves *Discovery* on 2 November 1902. The three-man party comprises Robert Scott, Edward Wilson, and Ernest Shackleton, with 19 dogs. Although they set a new furthest south of 82°16′S, they do not travel beyond the Ross Ice Shelf.

Jean-Baptiste Charcot's *Français* expedition sails to the western side of the Antarctic Peninsula, to rescue Nordenskjöld (who had disappeared) and carry out an ambitious program of surveying and science. The expeditioners return to France as polar heroes; the subsequent publication of 18 volumes of scientific reports amply justifies that standing.

The *Nimrod*'s Northern Sledging Party (Edgeworth David, Douglas Mawson, and Alistair Mackay), which set out from Cape Royds, reaches the South Magnetic Pole on 16 January. They claim it for the British Empire.

Background image: Norwegian explorer Roald Amundsen expressed some of the wonder of the Ross Ice Shelf: "Slowly it rose up out of the sea in all its imposing majesty. It is difficult ... to give any idea of the impression this mighty wall of ice makes on the observer who is confronted with it for the first time."

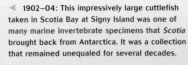

1902–04: This impressively large cuttlefish taken in Scotia Bay at Signy Island was one of many marine invertebrate specimens that *Scotia* brought back from Antarctica. It was a collection that remained unequaled for several decades.

1910–12: Members of the underfunded Japanese expedition showed the same resolve as many better known expeditions. The Japanese spent 60 hours cutting a path to the top of the Ross Ice Shelf before beginning a journey inland.

1910–12: The *Pourquoi-Pas?* was a much more comfortable place to spend the winter months than the *Français*, but Charcot wrote that only work and determination could save them "from being completely demoralized by the horrors of this climate."

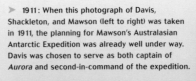

1911: When this photograph of Davis, Shackleton, and Mawson (left to right) was taken in 1911, the planning for Mawson's Australasian Antarctic Expedition was already well under way. Davis was chosen to serve as both captain of *Aurora* and second-in-command of the expedition.

1908–10: Jean-Baptiste Charcot (1867–1936) was born to wealth, but chose to spend much of his working life in harsh polar conditions. He always pointed out that Charcot Land and other features were named after his famous father.

1917: While trying to get the Ross Sea party ashore, Davis tried to push through thick ice to reach Knox Land.

1915: The strange device between Hurley (left) and Shackleton is a blubber-fuel stove that Hurley, who trained as a metal worker in his youth, made out of the *Endurance* ash chute and old oil drums. While resourceful, Hurley could be difficult, so the canny Shackleton picked him as a tent-mate to diffuse tension.

1903–05: When *Français* struck a submerged rock near Cape Tuxen, some of the crew tried to stem the leak, while others hand-pumped for 23 hours a day. Even as the ship finally sailed to safety, the crew held to their original goal and completed their mapping of the Palmer Archipelago.

1915: All 22 men slept under the two remaining boats upturned on rock walls 1 meter (3½ ft) high. Hurley wrote: "The 'Snuggery' grows more grimy day by day: everything is an oleaginous sooty black ... Oil mixed with reindeer hair, bits of meat, senegrass, and penguin feathers ..."

Nobu Shirase leads the Japanese South Polar Expedition. The expedition is dogged by vile weather and hostile, outright racist publicity. When their ship *Kainan Maru* puts in at Wellington in February 1911, the New Zealand press ridicules its "crew of gorillas" and claims that the polar regions are no place for "such beasts of the forests".

Douglas Mawson's Australasian Antarctic Expedition leaves Hobart on 2 December on board the *Aurora*, skippered by John King Davis. They establish a five-man radio station at Macquarie Island, and continue southward.

On 17 January, Robert Scott, Edward Wilson, "Birdie" Bowers, Edgar Evans and Lawrence "Titus" Oates reach the South Pole. They are dispirited to find the Norwegian flag flying there. All men perish on their wearisome return journey.

The remaining members of Mawson's expedition finally arrive in Adelaide on 26 February, after being forced to spend a second winter at Cape Denison.

Shackleton's Ross Sea Party is finally rescued (in the *Aurora*) after spending a perilous 199 days on the ice without adequate clothing or support. Three men lose their lives during the ordeal.

| 1910 | 1910–12 | 1911 | 1911 | 1911 | 1912 | 1913 | 1914 | 1915 | 1917 |

Norwegian explorer Roald Amundsen leaves Oslo on 7 June and heads for the South Pole rather than the North as he had announced.

In January, Robert Scott arrives at Ross Island, ready to launch his second assault on the South Pole. After enduring a bleak winter, Scott and the main polar party set out in November.

On 14 December, Roald Amundsen wins the race to the South Pole. He leaves a letter addressed to the King of Norway, telling of his accomplishments, and returns to his base at the Bay of Whales without incident.

A sledging expedition leads to death and disaster for the Australasian Antarctic Expedition. Mawson survives starvation, poisoning, blizzards and gaping crevasses, but his two companions, Dr Xavier Mertz and Belgrave Ninnis, perish. Alone, Mawson barely makes it back to the wintering hut at Cape Denison.

Ernest Shackleton's transcontinental expedition of 1914–17 spends a frightening 10 months on the ship *Endurance* when it becomes trapped in drifting ice. They abandon the ship in October and it is finally crushed and sinks in November. The expeditioners camp on the ice for five months before taking a hazardous boat journey to Elephant Island.

Expedition to the Ross Sea

After the Sixth International Geographical Congress, held in London in 1895, Sir Clements Markham, president of the Royal Geographical Society, wrote: "The exploration of the Antarctic regions is the greatest piece of geographical exploration yet to be undertaken." Three important scientific expeditions that flowed from that congress divided the frozen continent: the German trip led by Drygalski went to the region south of the Indian Ocean; the Swede Nordenskjöld traveled to the Antarctic Peninsula; and Britain focused on the Ross Sea.

The British National Antarctic Expedition

Markham recognized Robert Falcon Scott's potential when he first met the 18-year-old Royal Navy midshipman in 1887, and when Scott volunteered to lead the British National Antarctic Expedition in June 1899, Markham accepted his offer. The expedition combined the might of the Royal Navy, the Royal Geographical Society, and the Royal Society, and its aims combined exploration and science.

Their vessel was *Discovery*, the first ship designed and built in Britain for scientific exploration. It was constructed in Dundee of oak strengthened with internal beams and clad in a steel bow. *Discovery* sailed from the Isle of Wight on 6 August 1901, and called at Macquarie Island and Auckland Island on its way to Lyttelton, New Zealand, from where it sailed on 21 December. Crossing the Antarctic Circle on 3 January 1902, they reached Cape Adare six days later. Continuing into the Ross Sea, *Discovery* followed the edge of the Ross Ice Shelf eastward looking for the land reported by Ross, and eventually sighted the mountains of what is now Edward VII Land.

The expedition established its base at Hut Point on Ross Island and began to make some excursions. On one trip, Frank Wild displayed Antarctic ingenuity by hammering nails into the soles of his boots to provide extra traction on the ice. The Adélie penguins around the base were a source of endless fascination. Scott noted that Emperor penguins began to head south at the start of winter—the first recorded observation of their remarkable breeding pattern.

On 23 April the sun disap-

> ⟩ THEIR NAMES LIVE ON
> From left to right, the officers and scientific staff on the stern of the *Discovery* in 1901 are Wilson, Shackleton, Armitage, Barne, Koettlitz, Skelton, Scott, Royds, Bernacchi, Ferrar, and Hodgson. Many of their names are now immortalized as place names in Antarctica.

peared below the horizon, not to return until 22 August. The officers and men continued to live on *Discovery* and remained busy throughout the winter. For diversion there was a series of theatrical performances and the monthly *South Polar Times*.

In early October Charles Royds led a party to Cape Crozier where Reginald Skelton found Emperor penguins with well-developed chicks—a clear indication that they had hatched during the bitter Antarctic winter.

A sledging expedition

The first-ever attempt to sledge to the South Pole left *Discovery* on 2 November 1902. The three-man party comprised Scott himself, Edward Wilson, and Ernest Shackleton, with 19 dogs. Twelve men had left earlier to deposit supply caches for the polar party on their return journey. The groups soon met up and traveled together until 15 November, when the polar party

pressed on alone. Scott was confident at first, but progress was slow; the dogs were underfed and overloaded, and the men mostly had to haul the sledges themselves. Scott and Shackleton began to annoy each other. Years later, the amiable Wilson recalled an event on the ice cap. After breakfast, while he and Shackleton were loading the sledges, Scott shouted: "Come here you bloody fools." When Wilson asked if Scott was addressing him, Scott replied "No." "Then it must have been me," said Shackleton, and when Scott was silent, continued: "Right, you're the worst bloody fool of the lot, and every time you dare to speak to me like that, you'll get it back." It was an odd exchange in a place where each relied on the other for survival.

Although they set a new furthest south of 82°16′ on 30 December 1902, they did not travel beyond the Ross Ice Shelf. By then all were suffering from scurvy, and the last few dogs died on the return journey. They were back at Hut Point by 3 February 1903.

Homeward bound

Discovery was icebound so the expedition had to spend a second winter in Antarctica and undertook excursions over that winter and into the next summer. Two relief vessels arrived on 5 January 1904 to evacuate the party. They arrived in England on 10 September 1904.

> ◄ BIRD'S-EYE VIEW
> Scott was the first Antarctic balloonist, ascending to 244 meters (800 ft) in the basket below a hydrogen balloon. Shackleton, the only other one in the party to go up, took aerial photographs. Their height of ascent was limited by the weight of the heavy tethering rope.

▲ STARK REMINDER

Looking back from the tip of Hut Point across the Discovery hut to McMurdo Station, the cross in the foreground is a memorial to George Vince who died near this spot on 11 March 1902. On the first sledging expedition, Vince slid off an ice cliff into the sea, and his body was never found.

▲ PLACE OF REFUGE

The interior of the Discovery hut has been restored to look as it did when some of Shackleton's Ross Sea Party relied on it for their survival. The interior of the hut is stained black by smoke from the blubber that was their only fuel for heating, cooking, and lighting.

THE
SOUTH POLAR TIMES
1902 ⟷ 1903

THE ARMS OF THE "DISCOVERY."

◀ WHILING AWAY WINTER

During winter, a monthly journal was produced within the hut. Shackleton was elected editor of the South Polar Times. The magazine contained scientific reports, humor, and fiction, all contributed anonymously. This coat of arms was drawn by Edward Wilson, principal artist for both the magazine and the expedition.

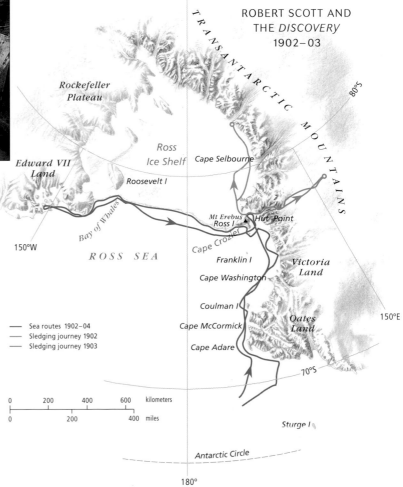

ROBERT SCOTT AND
THE DISCOVERY
1902–03

TRANSANTARCTIC MOUNTAINS

Rockefeller Plateau

Ross Ice Shelf

Edward VII Land

Cape Selbourne

Roosevelt I

80°S

Bay of Whales

Mt Erebus
Ross I

Hut Point

150°W

Cape Crozier

ROSS SEA

Franklin I

Victoria Land

Cape Washington

Coulman I

Oates Land

150°E

Cape McCormick

Cape Adare

70°S

— Sea routes 1902–04
— Sledging journey 1902
— Sledging journey 1903

0   200   400   600   kilometers
0   200   400   miles

Sturge I

Antarctic Circle

180°

Leader of men

When Ernest Shackleton arrived back in England on 12 June 1903, he found that Scott's 1901–04 expedition, from which he had been virtually sacked, was a controversial subject. Before departing, Scott had been told that the expedition was not to stay a second winter, and *Discovery* being icebound was the result of incompetence. Furthermore, Scott's expedition was broke, so the government would have to pay for the forthcoming relief voyage of *Morning* and *Terra Nova*. Shackleton, by contrast, was home and was a polar hero. He was soon declared fit for Antarctic service, and used his fame to raise his social standing. He also needed money. After a short period as a journalist he was appointed Secretary of the Royal Scottish Geographical Society, stood for Parliament in an election that he lost (along with his party), and set about organizing his own Antarctic expedition.

On 8 February 1907, William Beardmore, Shackleton's main sponsor, promised a large loan, and just three days later Shackleton was assembling his British Antarctic Expedition. However, he soon learned that Scott was planning to sail south again, and regarded not just the *Discovery* base but the whole of Ross Island as rightfully his; he insisted Shackleton should stay away. Pressured by their mutual colleague Edward Wilson, Shackleton agreed to this outrageous constraint.

Establishing winter quarters

The expedition left England on 7 August 1907 aboard *Nimrod*. The ship also carried a motorcar: an Arrol-Johnston made in Scotland at a factory owned by Beardmore. Shackleton left later, and met the ship in Lyttelton; *Nimrod* could only make six knots, so catching up was not hard. The 300-ton *Nimrod* was 40 years old and rather battered, but Shackleton could not afford his first choice, *Bjørn*. To save coal, *Nimrod* was towed from New Zealand into the ice by the larger *Koonya*, owned by New Zealand's Union Steamship Company and captained by Frederick Evans. The ships left New Zealand on New Year's Day 1908, and *Koonya* turned north just past the Antarctic Circle on 15 January after two weeks of life-threatening storms. Shackleton noted that *Koonya* was the first steel vessel to cross the Antarctic Circle.

At 9.30 am on 23 January, they sighted the Ross Barrier (now the Ross Ice Shelf) and sailed along it, looking for the bay in the ice that Scott had named

Balloon Bight after he and Shackleton went aloft there. They passed the bay where Borchgrevink had landed in 1900, "but it had greatly changed." Then they came to a large harbor that Shackleton named the Bay of Whales because "it was a veritable playground for these monsters" before he fled the heavy ice that was driving into this bay. When Shackleton realized that this was where he had ballooned, but the configuration had changed dramatically because a large section of the ice shelf had broken away, he resolved to make his winter base on land, not ice. When pack ice prevented him from continuing east, he broke his promise to Scott and turned west toward Ross Island. But even there sea ice stopped *Nimrod* from approaching the old Discovery hut at Hut Point, so they erected their prefabricated hut at Cape Royds. *Nimrod* left a wintering party of 15 when she sailed for Lyttelton on 22 February.

Overland excursions

Their first major land excursion was the first ascent of Mount Erebus. Edgeworth David (the leader), Douglas Mawson, Alistair Mackay, and a support party of three

left on 5 March, and struggled through savage blizzards to the summit (3,795 meters/12,450 ft) on 10 March; they were back two days later.

On 29 October, Shackleton, Frank Wild, Jameson Boyd Adams, and Eric Marshall set out for the South Geographic Pole. Shackleton planned to walk and use ponies, even though the advantages of skis and dogs for polar travel were by now widely recognized. The car, Antarctica's first, was virtually useless. By 26 November, Shackleton could again claim a record for the furthest south—further than he had penetrated with Scott—but he was worried that they were already rationing food.

◀ LACK OF FORESIGHT
The car garaged at Cape Royds was eventually returned to England. The device (right) is a maize crusher. Shipping a car but not learning to ski was typical of Shackleton—even on the *Endurance* expedition, six years later, he told Orde-Lees that he "had no idea how quickly it was possible for a man on skis to get about."

▼ SCALING A VOLCANO
All six men in the climbing party pushed on for the summit, but Brocklehurst suffered frostbite not far from the top. Here, at the summit, the men peer some 275 meters (900 ft) down into the active crater—about 1 kilometer (½ mile) wide—on one of the rare occasions when the steam cloud was swept aside by the wind.

◄ PONY POWER

Nansen, the greatest polar explorer of his time, personally advised Shackleton against the suitability of horses for drawing sledges in Antarctica. Even so, through the Hong Kong and Shanghai Bank, Shackleton ordered a dozen Manchurian ponies for his expedition.

► HOSTILE TERRAIN

The Taylor Glacier in the Dry Valleys was named by Douglas Mawson after geologist Griffith Taylor. While the long trudges across the Ice Shelf and over the polar plateau were feats of endurance, climbing the glaciers from one to the other was the biggest battle.

▼ RAISING THE FLAG FOR BRITAIN

Queen Alexandra gave the expedition a Union Jack, which the polar party erected at their furthest south—less than 160 kilometers (100 miles) from the Pole. Shackleton wrote: "While the Union Jack blew out stiffly in the icy gale that cut us to the bone, we looked south ... but could see nothing but the dead white snow plain."

ERNEST SHACKLETON AND THE *NIMROD* 1907–09

TRANSANTARCTIC MOUNTAINS

90°W

90°S
South Pole

90°E

Queen Alexander Range

Beardmore Glacier

Rockefeller Plateau

Shackleton Inlet

Ross Ice Shelf

Cape Selbourne

Edward VII Land

80°S

Roosevelt I

Bay of Whales

Bluff Depot

Cape Royds

Ross I

ROSS SEA

Franklin

Cape Washington

Drygalski Ice Tongue

Victoria Land

150°W

Coulman I

Oates Land

George V Land

Cape McCormick

Cape Adare

70°S

150°E

Cape Hudson

— Shackleton's journey south
— Magnetic polar party
— Sea route 1908
— Sea route 1909

Sturge I   Balleny Islands

Scott I

Young I

Antarctic Circle

180°

0   200   400   600   kilometers
0   200   400   miles

In early December they became the first men to reach the southern extent of the Ross Ice Shelf and to climb the vast Beardmore Glacier onto the polar ice cap. The climb was a nightmare, and on 7 December the last pony fell into a crevasse and died. Just after Christmas, the slope became easier and they were on the ice cap, but they had only one month's food to reach the Pole and return to their supply dump—a journey of almost 800 kilometers (500 miles). They were all suffering from altitude sickness, and Shackleton described the symptoms: "My head is very bad ... as though the nerves were being twisted up with a corkscrew and then pulled out." He was torn between his ambition to reach the Pole and his sense of responsibility for his men. By 2 January he was writing: "I must ... consider the lives of those who are with me. I feel that if we go on too far it will be impossible to get back ... man can only do his best ... time is going on and food is going also." On 9 January they left everything behind and made a one-day push to the south—to come within 160 kilometers (100 miles) of the Pole—and back. At 88°23′S they hoisted the Union Jack, took some photographs, and left a commemorative brass cylinder. Looking southward, Shackleton noted: "We feel sure that the goal we have failed to reach lies on this plain ... we stayed only a few minutes ...

Whatever regrets may be, we have done our best." By the time they returned to their camp that day, they had traveled 66 kilometers (41 miles).

On the return journey, they hoisted a sail on their sledge and quickly reached the top of the Beardmore Glacier, but descending to their food dump while weak and short of food was extremely draining: "I cannot describe adequately the mental and physical strain of the last 48 hours," wrote Shackleton of 26 and 27 January.

A narrow escape

They were near death, yet on 31 January Shackleton surreptitiously gave Frank Wild his breakfast biscuit. Wild wrote: "I do not suppose that anyone else in the world can thoroughly realize how much generosity and sympathy was shown by this; I do by God I shall never forget it." They all had acute dysentery and were "appallingly hungry," and by 21 February Shackleton was writing: "Our need is extreme and we must keep going … food lies ahead and death stalks us from behind." Two days later they arrived at Bluff Depot and food supplies, but Marshall was so ill that Shackleton and Wild left him with Adams and pressed on to arrange rescue. They found the building at Hut Point empty, but there was a note saying that the ship was nearby, so they set fire to a hut to attract attention. It worked; by

1 March both were on board the *Nimrod*. After "a good feed of bacon and fried bread," the indomitable Shackleton set off with three companions on the 18½ hour trek to rescue Adams and Marshall. Shackleton wrote: "We were all safe on board at 1 am on March 4." The polar party had walked 2,740 kilometers (1,700 miles) in 128 days.

Looking for the South Magnetic Pole

Meanwhile, the others had also had some harrowing experiences. Their most ambitious expedition was the Northern Sledging Party to the South Magnetic Pole, with a party of three: Edgeworth David, Douglas Mawson, and Alistair Mackay. David and Mawson were Australian geologists, and Mackay was a naval surgeon. David, the leader, turned 51 during the journey, Mawson was 26, and Mackay 30. They tried using the car to transport supplies across the sea ice, but soon abandoned it in favor of man-hauling—a challenging task, as one of the two sledges weighed 275 kilograms (606 lb) and they had some 1,600 kilometers (1,000 miles) to cover.

▲ FAILED HORSEPOWER
The motor car was adequate on hard ice but failed in the least depth of snow. On 22 September the car hauled sledges for eight miles on a depot-laying exercise before snow forced the men to take up the yokes themselves.

▲ MAGNIFICENT STAMINA
Tannatt William Edgeworth
David (1858–1934) turned 51
during the *Nimrod* expedition.
Nevertheless, he climbed to
the summit of Mount Erebus
and completed the difficult
2,028-kilometer (1,260-mile)
walk to the South Magnetic
Pole. He was knighted in 1920.

▼ SECOND PROUD MOMENT
The Magnetic Pole party (from
left: Mackay, David, Mawson)
claimed the area at 3.30 pm
on 16 January 1909. Mackay
and David set up the Union
Jack while Mawson positioned
the camera. David recalled
that he pulled the string to
trigger the camera: "Then we
gave three cheers for His
Majesty the King."

They left Cape Royds on 5 October, while supply
depots were still being set up for the geographic polar
bid. Shackleton was to cross the ice shelf while the
sledging party walked northward on the sea ice along
the edge of Victoria Land (which they claimed for
Britain on 17 October), with Mawson as pathfinder.
Heavy loads and soft snow slowed them down, so
they cached some of their rations and relied on seal
and penguin meat, cooked on an improvised stove.
They could not abandon any sleeping gear; they
already shared a three-man sleeping bag and kept
each other awake. Their intended route up the
Drygalski Ice Tongue proved impractical. David
recalled that, by 20 December, "We had not yet climbed
more than 100 feet [36 meters] or so above sea level …
We knew that we had to travel at least 480 to 500 miles
[770 to 800 kilometers] to the Magnetic Pole and back
to our depot, and there remained only six weeks to
accomplish this journey." They pushed on with a single
sledge weighing 305 kilograms (670 lb), and climbed a
route they named Backstairs Passage.

By 11 January 1909 they were on the plateau and
more than 3,800 meters (7,000 ft) above sea level. Their
compass was only 15 degrees off vertical, but that did
not mean that they had only 15 nautical miles
(equivalent to about 28 kilometers) more to travel, as
they had to allow for the daily movement of the
Magnetic Pole, and might even have to chase it.
According to David, "Mawson considered that we were
now practically at the Magnetic Pole, and that if we
were to wait for twenty-four hours taking constant
observations at this spot the Pole would, probably …
come vertically beneath us." But rather than wait for
the Pole to come to them, they decided to go to its
approximate mean position. Like Shackleton, a week
earlier and much further to the geographic south, they
left everything and made a one-day rush for their goal
on 16 January. They hoisted the Union Jack, claimed
the area for the British Empire, gave three cheers for the
King, and took a photograph. Then, "with a fervent
'Thank God' we all did a
right about turn, and as
quick a march as tired limbs
would allow back in the
direction of our little green
tent in the wilderness of
the snow."

Several years later,
calculations from Mawson's
observations showed that
the party had been close to
the area of polar oscillation
but had not penetrated it.
When Mawson learned of
this in 1925, he amended
his entry in *Who's Who in
Australia* from "one of
the discoverers of the
South Magnetic Pole"
to "Magnetic Pole
journey 1908."

A fortunate rescue

To return to meet *Nimrod* at their depot on the Drygalski
Ice Tongue, they had to average 27 kilometers (17 miles)
for 15 days. They had full rations, but were short of tea
and took to recycling tea bags that they had discarded
on the outward journey and salvaged as they went back.
They reached the depot on the morning of 3 February;
only a few hours later they heard a rocket, and rushed
from their tent to see *Nimrod* very close. A distinguished
career was almost snuffed out when Mawson fell into a
crevasse and had to be rescued, but then a journey of
2,030 kilometers (1,260 miles) over 122 days ended
luxuriously with their first baths in four months.

Only later did they learn how lucky they were
to be found. *Nimrod* had a daunting 320 kilometers (200 miles) of coast to search for their three
small figures. John King Davis, First Officer, was
on duty from 4 am to 8 am on 3 February. When
Captain Evans came on deck before breakfast and
confided that he thought that there was little
chance of finding the land party, he asked Davis
if his watch had examined the entire coast. Davis
admitted that a small section had been obscured
by icebergs. After balancing the extra fuel expen-
diture against the likelihood of the men being in
just that location, they decided to steam back for
four hours. Sailing into a narrow fault in the ice
behind the obstructing icebergs, they saw that
"upon the crest of a little knoll of ice was a green
conical tent." When *Nimrod* had first passed this
spot the expeditioners had been further up the
glacier. If Davis had not missed it and decided to
return, the Magnetic Pole party would have been
left to make their own way back to Cape Royds.
Much of the sea ice had broken up, and their
chances of surviving would have been slight.

With David, Mackay, and Mawson on board,
Nimrod returned to Cape Royds, and then to
Hut Point where they found Shackleton and his
party. On the way back to New Zealand,
Shackleton sailed along the coast beyond Cape
Adare to map the coast of Adélie Land as far west
as possible, and reached 166°14′E—beyond any
previous effort. They reached Lyttelton on
25 March and were back in England in June.

Shackleton was knighted for his achievements, but
he thought he had failed, although he had paved the
way to the Pole. Roald Amundsen was later to write
of Shackleton: "Seldom has a man enjoyed a greater
triumph; seldom has a man deserved it better." More
specifically, he said: "I admire in the highest degree
what [Shackleton] and his companions achieved with
the equipment they had. Bravery, determination,
strength they did not lack. A little more experience …
would have crowned their success." Amundsen also
wrote what could well be the defining judgment of
Shackleton: "Sir Ernest Shackleton's name will for
evermore be engraved with letters of fire in the history
of Antarctic exploration. Courage and willpower can
make miracles. I know of no better example than what
that man has accomplished."

> LIONIZED BY THE PRESS

After the *Nimrod* expedition, the *Daily Telegraph* gushed: "In his photographs Mr. Shackleton is nothing more than an intelligent but ordinary naval officer. In reality he radiates the fascination of an indefinable force ... If he has the face of a fighter, he has the look of a poet; one must be both a fighter and poet to accomplish what he has done."

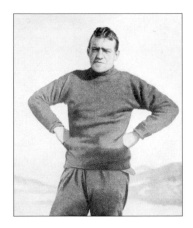

⌃ LOOKING TO THE FUTURE

When Shackleton left Cape Royds in 1909, he noted: "I left ... stores sufficient to last fifteen men one year. The vicissitudes of life in the Antarctic are such that such a supply might prove of the greatest value to some future expedition."

▽ FOOTWEAR FOR POLAR CONDITIONS

Shackleton brought both ski boots and finnesko with him. Finnesko are boots of reindeer fur filled with sennegrass that both adds insulation and absorbs moisture so feet stay warm and dry.

⌃ STANDING STILL

The hut at Cape Royds has changed little. A wheel for the motor car still lies near the collapsed garage. However, there has been considerable restoration— visitors in 1947 found boards off the roof of the hut and the garage had already collapsed. The roof was repaired and windows were re-glazed on later visits.

> TIME CAPSULE

The hut was restored by a New Zealand team in 1961 to look as it did in 1909. Shackleton wrote that the cooking range was designed "to burn anthracite coal continuously day and night and to heat a large superficial area of outer plate, so that there might be plenty of warmth given off in the hut."

Norway's bid for the South Pole

▲ HIGH ACHIEVER
It seems likely that Roald
Amundsen (1872–1928) was
not only the first to walk to
the South Pole, but may also
have reached the North Pole
first (in an airship), as well as
being the first to sail through
the Northwest Passage.

▼ CUSTOM DESIGNED
Fram (meaning "Forward") was
built with a rounded hull so
that the pressure of ice would
cause her to rise, rather than
be crushed. The ship was
launched in 1892, and had
already performed well in the
Arctic before Nansen lent her
to Amundsen.

Amundsen has been described as a consummate professional in an age of amateurs. The story of his journey to the South Pole certainly reveals organizational skills that his competitors lacked, but there were also controversial elements, even errors, in the Norwegian polar bid.

Amundsen's chronicle of his expedition reveals a warmhearted, generous personality, far from the cunning manipulator that Scott's supporters painted him at the time. He rejected the title of adventurer: "For him [the explorer], adventure is only an unwanted interruption … He does not seek titillation, rather previously unknown facts … an adventure is simply an error in his calculations … Or is it a very unhappy proof that no one can take into consideration all eventualities? … Any explorer experiences adventures. They excite him and he looks back on them with pleasure, but he never seeks them out."

A change of direction

Amundsen had been the first to conquer the Northwest Passage, and had spent a winter in Antarctica with Adrien de Gerlache on *Belgica*. He was a meticulous planner, but in 1909, when he was aiming to be first to the North Pole, and had borrowed Fridtjof Nansen's *Fram* for the purpose, events forestalled him: in September of that year, both Robert Peary and Frederick Cook claimed the North Pole. Amundsen recalled: "Just as rapidly as the message had traveled over the cables I decided on my change of front—to turn to the right-about, and face to the South." With a secrecy that later attracted criticism, he told only his brother, and later Lieutenant Thorvald Nilsen, captain of *Fram*. The crew found out in Madeira; until then, they believed they were sailing around Cape Horn and up to the Bering

Strait to take advantage of the Arctic drift. They all agreed to stay for the new destination.

Fram left Christiania (now Oslo) on 7 June 1910, and sailed from Madeira on 9 September, with 19 men and 97 dogs—the number of dogs had risen to 116 by January 1911. Amundsen telegraphed Scott, stating: "Beg leave to inform you *Fram* proceeding Antarctic. Amundsen," and notified his sponsors and Nansen in the same way. Secrecy was not Amundsen's only odd trait: he also would not take a doctor on his expeditions, largely because he thought an expedition doctor, being well educated, could encourage dissent.

Having read reports of previous expeditions, Amundsen decided to establish his base on the ice at the Bay of Whales because it was closer to the Pole than Ross Island. *Fram* arrived there on 14 January, and the crew erected their prefabricated hut at "Framheim." They killed seals and penguins for food, and Amundsen wrote: "An Emperor penguin just came on a visit—soup-kettle." On 4 February, Scott's ship *Terra Nova* visited for a day, but Scott had already been left on Ross Island. Amundsen's men established supply depots as far as 82°S, and *Fram* departed northward before the Bay of Whales froze over.

Surprisingly, on an expedition so well planned, there was one key blunder: they had the Almanac for 1911, but had forgotten the Almanac for 1912, which contained essential information for fixing the position of the Pole. If Amundsen were to claim the Pole, he had to be there before the end of 1911.

◄ FOUR-LEGGED FRIENDS
On the voyage south, Rønne the
sailmaker danced with one of the
dogs—"in the absence of lady
partners"—to a tune on the fiddle.
Amundsen also reported that
kenneling was adjusted to house
dogs that were friendly with one
another side by side, and they
would be in the same teams.

When the sun appeared on 24 August after four months of darkness, the sledges were already packed for the polar run. Indeed, Amundsen set off for the South Pole in September—very early in spring—but had to return because conditions were too cold for survival. The dogs, after six months of leisure, did not "understand that a new era of toil had begun." On 19 October five men started for the Pole on skis and sledges, and by 7 November they were far enough across the Ross Ice Shelf to see mountains to the south. At the base of the climb to the polar ice cap, Amundsen outlined his plan: "The distance … from this spot to the Pole and back, was 683 miles [1,100 km]… we decided to take provisions and equipment for sixty days on the sledges, and to

▲ ONE-SHACK TOWN

"Framheim" in February 1911 was an orderly place. *Fram* was about to depart and the first of the depot laying parties was ready to set out. Some 900 cases had been unloaded and the hut was built. Its walls were tarred and the roof covered with tarred paper.

▼ STRATEGIC POSITIONING

Much of Amundsen's success came from his choice of sites. The Bay of Whales, seen here in early morning, was one degree closer to the Pole than McMurdo Sound. Shackleton had turned away from there fearing that the Barrier ice could calve and leave his whole camp adrift.

▲ THE ART OF SKIING
Skis were heavy and a single pole was used. Amundsen had been skiing since a schoolboy in Norway; his first long trip was an 80-kilometer (50-mile) journey that took 20 hours.

▶ EXERCISING THE DOGS
This image of the Norwegian camp at 85°S first appeared in the *London Illustrated News*. Amundsen commented: "We had thought that a day's rest would be needed by the dogs for every degree of latitude but ... instead of losing strength [they became] ... more active every day."

leave the remaining supplies ... and outfit in depot ... We now had forty-two dogs. Our plan was to take all forty-two to the plateau; there twenty-four of them were to be slaughtered, and the journey continued with three sledges and eighteen dogs. Of the last eighteen, it would be necessary, in our opinion, to slaughter six in order to bring the other twelve back to this point. As the number of dogs grew less, the sledges would become lighter and lighter, and when the time came for reducing their number to twelve, we should only have two sledges left."

His calculations were almost perfect: they took eight days less than the time allowed, and returned with 12 dogs. They climbed the Axel Heiberg Glacier from 17 to 21 November, and with some sadness, because a "trusty servant lost his life each time;" they shot the dogs at a camp they called the Butcher's Shop at the top of the glacier.

On 8 December they passed Shackleton's 88°23′S. It was a milestone for the 39-year-old Amundsen, who wrote: "No other moment of the whole trip affected me like this. The tears forced their way to my eyes ..." They laid down their final depot here, and made a push for the South Pole.

Arriving at the South Pole

Of the events of 14 December 1911, Amundsen wrote:

"At three in the afternoon a simultaneous "Halt!" rang out from the drivers. They had carefully examined their sledge-meters and they all showed the full distance—our Pole by reckoning. The goal was reached, our journey ended. ... I had better be honest and admit straight out that I have never known any man to be placed in such a diametrically opposed position to the goal of his desires as I was at that moment. The regions around the North Pole—well, yes, the North Pole itself—had attracted me from childhood, and here I was at the South Pole. Can anything more topsy-turvy be imagined?

"... Pride and affection shone in the five pairs of eyes that gazed upon the flag, as it unfurled itself with a sharp crack, and waved over the Pole. I had determined that the act of planting it—the historic event—should be equally divided among us all. It was not for one man to do this; it was for *all* who had staked their lives in the struggle, and held together through thick and thin.

"... Everyday life began again at once ... Of course, there was a festivity in the tent that evening—not that champagne corks were popping and wine flowing—no, we contented ourselves with a little piece of seal meat each, and it tasted well and did us good."

With the exactness that characterized the whole expedition, they spent the next few days surveying around their camp until they were certain that they had reached the Pole by any calculations. On 17 December, at the point they concluded was the Pole, they erected a tent with a Norwegian flag on top and "inside the tent, in a little bag, I left a letter, addressed to HM the King, giving information of what we had accomplished. The way home was a long one, and so many things might happen ..." They also left some clothes and a sextant, before lacing the tent and turning to the north.

The secret of success

By 6 January they were back on the ice shelf, and at 4 am on 25 January they reached Framheim, after a 99-day journey. They were probably fitter and healthier than when they left. The science of nutrition was in its infancy, but at Framheim the Norwegians ate a healthy mixture of wholemeal bread, berry preserves, and undercooked seal meat. Whereas Shackleton's party barely struggled back to rescue and Scott's party did not come back at all, because their rations resulted in starvation as well as nutritional deficiencies, Amundsen's

90°W

90°S *South Pole*
14 Dec 1911

90°E

T R A N S A N T A R C T I C   M O U N T A I N S

ROALD AMUNDSEN
REACHES THE
SOUTH POLE 1911

*Axel
Heiberg
Glacier*

*Rockefeller
Plateau*

*Ross
Ice Shelf*

Cape Selbourne
80°S

*Edward VII
Land*

Roosevelt I

Bay of Whales

Mt Erebus
Ross I

150°E

R O S S   S E A

150°W

| 0 | 200 | 400 | 600 | kilometers |
| 0 | | 200 | | 400   miles |

180°

men ate well, and had so much food on hand that he could write: "We are bringing the purveyors of our sledging samples of their goods that have made the journey to the South Pole and back in gratitude for the kind assistance they afforded us." Shackleton sacrificed a biscuit; Amundsen brought back food as souvenirs.

Five days after they arrived back at their coastal base, the men were sailing to Hobart on *Fram*. On 7 March 1912, Amundsen telegraphed news of his success to his brother, who announced it to the world. When Amundsen heard that Mawson needed dogs for his Australasian Antarctic Expedition; he donated 21, keeping only puppies and survivors from the conquest of the South Pole.

Scott's fateful journey

Scott arrived in Melbourne on 12 October 1910 on his way to Antarctica to unwelcome news: Roald Amundsen was also bound for the South Pole. The race had begun. As well as fretting about the unexpected personal competition, Scott was annoyed that the Norwegians were aiming for the Pole that he considered his by right. Science was important to his second polar expedition, but so was claiming the Pole for Britain. However, from the start Scott doubted that he would win the race; on 11 October, he wrote: "I don't know what to think of Amundsen's chances. If he gets to the Pole, it must be before we do, as he is bound to travel fast with dogs and pretty certain to start early."

Scott had announced his second Antarctic expedition in September 1909, but money was slow to come in, despite widespread endorsement. The *Terra Nova*, which had come to his support on the previous expedition, was to be the expedition ship. Edward "Teddy" Evans was to be his second-in-command. Lawrence "Titus" Oates and Apsley Cherry-Garrard each contributed a thousand pounds, and were recruited as an officer and assistant zoologist, respectively. Oates wrote to his mother: "Points in favor of going. It will help me professionally as in the army if they want a man to wash labels off bottles they would sooner employ a man who has been to the North Pole [*sic*] than one who has only got as far as the Mile End Road. The job is most suitable to my tastes. Scott is almost certain to get to the Pole and it is something to say you were with the first party. The climate is very healthy although inclined to be cold."

On the way to the Pole

Terra Nova left Britain at the beginning of June 1910, and Scott joined the ship in South Africa. The expedition photographer, Herbert Ponting, boarded in Lyttelton, New Zealand, with 19 Siberian ponies and 33 sledge dogs (and one collie bitch) shipped from Russia. After three weeks in the ice, the ship was off Ross Island on 4 January 1911. Hut Point and Cape Crozier were frozen in, so they erected their hut at Cape Evans. It had a wardroom for the 16 officers and scientists and a separate mess deck for the nine crewmen.

Before winter, a western party led by Griffith Taylor went to the Dry Valleys, and an eastern party sailed *Terra Nova* toward King Edward VII Land, and soon returned to tell of meeting Amundsen and the *Fram*, so the land party it carried had to move to Cape Adare. Scott was surprised to hear that the Norwegians were so close; he had assumed that they would start from the Weddell Sea side of the continent. He wrote: "There is no doubt that Amundsen's plan is a very serious threat to ours. He has a shorter distance to the Pole by 60 miles [97 kilometers]—I never thought he could have got so many dogs safely to the ice. His plan for running them seems excellent. But above and beyond all he can start his journey early in the season—an impossible condition with ponies."

Scott knew the limitations of his ponies. He had led the southern party intending to set up a supply depot at 80°S, but while the 26 dogs performed well, the eight ponies kept sinking in the soft snow. Even when the party switched to marching at night, when

▲ A HOME AT CAPE EVANS
Beyond the Cape Evans hut stands the Greenpeace Antarctic
base (now removed) and beyond that, Mount Erebus. The hut's
interior appearance now owes as much to Shackleton's later
Ross Sea Party, which was marooned here and lived off Scott's
excess rations, as it does to the Scott expedition itself.

▶ APPRECIATING THE RETURN OF DAYLIGHT
One of the expeditioners looks across the hut and
the stores to Mount Erebus. The return of the sun in
spring was a major event. Scott wrote: "It changes
the outlook on life of every individual, foul weather is
robbed of its terrors; if it is stormy today it will be
fine tomorrow or the
next day, and each day's
delay will mean a
brighter outlook when
the sky is clear."

◀ SCOTT'S SANCTUM AT CAPE EVANS
Scott, often uncomfortable in his dealings with people,
was a skilled writer with a versatile mind. In the hut he
appeared to be happiest reading and writing at his desk
in his den, or presiding over academic discussions on a
wide range of topics at the large oak table.

◄ CROWDED HOUSE

CROWDED HOUSE

The hut had a big oak table for the officers and a smaller table for the men. The larger one (bearing a guest book) is still in the hut today. Herbert Ponting's darkroom is at the end of the room and the Tenements are on the left.

▼ NEGATIVE SPACE

Ponting's darkroom, at the hut's rear, today resembles a historical display of the early days of photography. Besides a tripod, there are chemicals and developing tanks and scientific equipment used in various experiments.

▲ CHEEK BY JOWL

The men lived in very close proximity in the Tenements. From left to right: Cherry-Garrard, Bowers, Oates, Meares, and Atkinson. Scott noted on 17 January 1911: "I saw Bowers making cubicles as I had arranged, but I soon saw these would not fit in, so instructed him to build a bulk-head of cases which shuts off the officer's space from the men's, I am quite sure to the satisfaction of both."

the snow was firmer, progress was slow. Finally, One Ton Camp was established 58 kilometers (equivalent to about 31½ nautical miles) short of their goal—a short-fall that proved fatal.

Scott insisted on a regular routine through the dark winter months. The most remarkable exploit was the bleak excursion recounted in Apsley Cherry-Garrard's *The Worst Journey in the World*. With Edward "Bill" Wilson and "Birdie" Bowers, he set out in the dark and bitter cold for Cape Crozier, 100 kilometers (62 miles) away to collect Emperor penguin eggs because Wilson believed they might provide a clue to the evolution of reptiles into birds. They returned five weeks later, with just three eggs, and ravaged faces. (The eggs were analyzed years later, with inconclusive results.) Cherry-Garrard's masterpiece of polar literature begins: "Polar exploration is at once the cleanest and most isolated way of having a bad time which has been devised," and ends: "if you have the desire for knowledge go out and explore … If you march your Winter Journeys you will have your reward, as long as all you want is a penguin's egg."

The assault on the Pole

On 24 October 1911 the motor sledges set off to lay depots, and on 1 November the main polar party, led by

▲ PICTURE SHOW

On Monday 29 May 1911 Herbert Ponting gave a lecture on Japan using his own collection of illustrations. Scott noted: "He is happiest in his descriptions of the artistic side of the people, with which he is in fullest sympathy."

Scott, left Cape Evans. The two groups comprised 16 men, 10 ponies, 33 dogs, and 15 sledges, including the two motorized ones. By 6 November the polar party had passed the abandoned motor sledges—their crew had continued south but were now man-hauling. On 21 November the two groups met, and on 9 December they reached the bottom of the Beardmore Glacier: Scott was literally following in Shackleton's footsteps. Here they shot the surviving ponies, and men and dogs carried on up the glacier. It was hard work, especially as they could not ski, and even Scott admitted that "it would be impossible to drag sledges on foot." On 11 December the dogs were sent back and 12 men continued, hauling three heavily laden sledges. On 14 December Amundsen reached the South Pole while Scott was still on the Beardmore Glacier, writing: "We got bogged again and again, and, do what we would, the sledge dragged like lead … Considering all things, we are getting better on

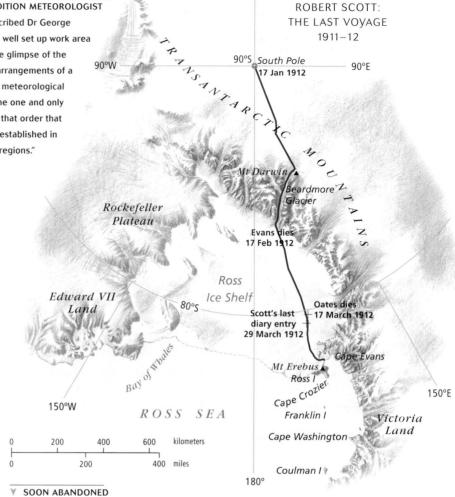

◄ EXPEDITION METEOROLOGIST
Scott described Dr George Simpson's well set up work area as "A mere glimpse of the intricate arrangements of a first-class meteorological station—the one and only station of that order that has been established in the polar regions."

ROBERT SCOTT:
THE LAST VOYAGE
1911–12

skis." By 21 December Amundsen was on his way home and Scott's party was at Upper Glacier Depot. One of the supporting sledge teams turned back here, and eight men continued with two sledges.

On 3 January Scott made a fatal decision. He told his last support team to turn back the next day, but that one of them, "Birdie" Bowers, would come with him to the Pole. The sledge was already fully laden with supplies for four and Bowers had no skis, so the team would be limited to his pace. Even so, on Thursday 4 January Tom Crean, "Teddy" Evans, and William Lashly headed back, while Scott, Wilson, Oates, Bowers, and Petty Officer Edgar Evans carried on—never to be seen alive again.

At first things went well. On 9 January they passed Shackleton's furthest south (88°23′S), and Scott wrote: "All is new ahead." But by 11 January he was expressing doubts: "We ought to do the trick, but oh! for a better surface." On 16 January they saw a cairn of Amundsen's; Scott's frustration is clear: "The worst has happened … sledge tracks and ski tracks going and coming and the clear trace of dog's paws … The Norwegians have forestalled us and are first at the Pole … All the day dreams must go; it will be a wearisome return. Certainly we are descending in altitude—certainly also the Norwegians found an easy way up."

They reached what they thought was the Pole the next day, and found the Norwegians' tent the day after. Scott wrote despairingly: "The Pole. Yes, but under very different circumstances from those expected. We have had a horrible day … Great God! this is an awful place and terrible enough for us to have labored to it without the reward of priority … Now for the run home and a desperate struggle."

▼ SOON ABANDONED
The Wolseley four-cylinder air-cooled motor sledge, operating with caterpillar track, was an idea ahead of its time, and a precursor to the oversnow vehicles of today. However, it was poorly made and came with inadequate spare parts.

▲ PRECISION INSTRUMENT
Lieutenant "Teddy" Evans surveying with the four-inch theodolite that was used to locate the South Pole. Evans had been considering leading his own expedition to Antarctica to explore what is now the Edward VII Peninsula, but Scott adopted the plan as his own and offered Evans the role of his second-in-command.

► EXHAUSTED AND WEATHERWORN
Ponting took this photograph soon after "Birdie" Bowers, Edward "Bill" Wilson, and Apsley Cherry-Garrard (left to right) returned from "the worst journey in the world." Cherry-Garrard was the only one of the trio who returned to England, where the three eggs they had gathered were finally analyzed.

The way back

The dispirited party started back, but by 23 January Scott reported that Evans, the strongest of them, "is a good deal run down," and that "Oates gets cold feet." They reached Upper Glacier Depot on 7 February. Although they had mislaid or misjudged a day's supply of biscuits, and although Scott had noticed that Evans was "going steadily downhill," they spent two days adding 16 kilograms (35 pounds) of mineral specimens collected at Mount Darwin to their sledging load. Scott noted: "The morraine was obviously so interesting that … I decided to camp and spend the rest of the day geologizing. It has been extremely interesting." The fossils that the rocks contained later proved valuable in assessing the geological history of Antarctica, but the time spent collecting them and the effort of carrying them contributed to their deaths.

On Saturday 17 February, near the foot of the Beardmore Glacier, Edgar Evans lost his life. He had had several bad falls, and was straggling behind, and then disappeared and was found partly undressed and delirious. He sank into a coma and died that night.

By early March they were all in poor condition, particularly Oates, who had severe frostbite in his feet. On 10 March Scott wrote: "Things steadily downhill. Oates' foot worse. He has rare pluck and must know that he can never get through. He asked Wilson if he had a chance this morning and of course Bill had to say he didn't know. In point of fact he has none. Apart from him, if he went under now, I doubt whether we could get through. With great care we might have a dog's chance, but no more."

A week later Scott wrote: "Friday March 16 or Saturday 17—Lost track of dates but think the last correct … At lunch the day before yesterday, poor Titus Oates said he couldn't go on: he proposed we should leave him in his sleeping bag. That we could not do." They struggled on, but the same entry records: "He slept through the night before last, hoping not to wake; but he woke in the morning—yesterday. It was blowing a blizzard. He said 'I am just going outside and may be some time.' He went out into the blizzard and we have not seen him since … I take this opportunity of saying that we have stuck to our sick companions to the last."

Scott's other expeditioners

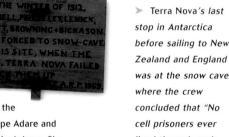

Often overlooked in accounts of Scott's expedition is the Northern Party, renamed the Eastern Party when, after failing to find a landing place on the eastern side of the Ross Sea, it moved to Cape Adare and erected a hut near the Borchgrevink expedition's base. Six men, led by Victor Campbell, landed from *Terra Nova* on 18 February 1911. Bad snow conditions greatly limited their explorations before they were picked up by *Terra Nova* almost a year later, on 3 January 1912. They decided to explore further along the coast of Victoria Land, starting at Terra Nova Bay on 8 January. They were due to be picked up around 18 February, and had ample supplies for their intended stay of about six weeks. However, the ship could not get back to them, so they had to dig a small snow cave on Inexpressible Island and spend the winter there. In poor health after a winter diet of penguins and seals, they left their cave at the end of September 1912 and traveled along the sea ice to Ross Island, surviving only because they found some food caches along the way. They reached Hut Point on 6 November, and learned of the fate of Scott's party from a note reporting that the polar party had not returned by the start of winter. They were recuperating at Cape Evans on 25 November when the search party returned with the news that they had found the bodies of Scott, Wilson, and Bowers.

> *Terra Nova's last stop in Antarctica before sailing to New Zealand and England was at the snow cave where the crew concluded that "No cell prisoners ever lived through such discomfort."*

> *The Northern Party first named the 12-kilometer (7-mile) long island "Southern Foothills" but later called it Inexpressible Island after living in the tiny snow cave.*

When they finally reached the South Pole, Scott's party discovered that the Norwegians had been there before them. Bitter disappointment is etched on their faces after performing the empty formalities that they had expected to be a rite of victory. From left to right: Wilson, Evans, Scott, Oates, and Bowers.

On March 19 a blizzard held them down in their tent just 11 miles from One Ton Depot, and their fuel ran out three days later. Scott's diary entry for 29 March records his last words: "Outside the door of the tent it remains a scene of whirling drift. I do not think we can hope for any better things now. We shall stick it out to the end, but we are getting weaker … and the end cannot be far.

"It seems a pity but I do not think I can write more—

"R. Scott

"Last entry

"For God's sake look after our people."

Aftermath

Eight months later, on 12 November 1912, expedition survivors found their tent and their bodies. The tent was closed and they erected a cairn of ice above it, with skis forming a rough cross on top. They marched 20 miles south to look for the body of Oates, but found the area covered in snow so built a cairn in his memory. Finally, they erected a huge wooden cross on the summit of Ross Island's Observation Hill, with a dedication that concluded: "to strive, to seek, to find, and not to yield."

Judgment of the British Antarctic explorers of the Heroic Age has shifted in prevailing attitudes. Scott was cast as a hero for most of the twentieth century, but recent critics have suggested that his only achievements were to write well and to die well. Shackleton has even been called unpatriotic because in 1909 he had enough provisions to reach the Pole but turned back. He and his companions could have claimed the Pole for Britain and died on their way back. By not sacrificing himself and his men, Shackleton is held to have left it to a foreigner to be first to the Pole. Scott was unlikely to have made the same decision. On the contrary, his message to the public as he was dying ran, in part: "The causes of the disaster are not due to faulty organization but to misfortune in all risks which had to be undertaken … I do not regret this journey, which has shown that Englishmen can endure hardships, help one another and meet death with as great a fortitude as ever in the past." Scott's memorial service in St Paul's Cathedral in February 1913 was attended by King George V, and Scott's widow, Kathleen, received his knighthood.

▲ SCOTT'S FINAL WORDS
The last entry in Scott's diary was taken as a testament to British spirit. Huntford, Scott's biographer, states that the Admiralty even announced that Scott and his men were considered as having been "killed in action" and a naval sermon was given on "the glory of self-sacrifice, the blessing of failure."

▲ TRIBUTE TO HEROIC FAILURE
A 2.7-meter (9-ft) memorial cross, made from tough Australian jarrah, stands on the summit of Observation Hill. It lists the men's names, and is further inscribed: "Who died on their return from the Pole. March 1912. To strive, to seek, to find, and not to yield." The words by Lord Tennyson were Cherry-Garrard's suggestion.

A dedicated explorer

After Shackleton's *Nimrod* expedition, Edgeworth David said: "Mawson was the real leader and was the soul of our party to the magnetic pole. We really have in him an Australian Nansen, of infinite resource, splendid physique, astonishing indifference to frost." Both Shackleton and Scott asked the Australian explorer and geologist to join their next expeditions. Mawson turned Scott down when Scott rejected his proposal to explore the coast west of Cape Adare. Shackleton and Mawson developed an extensive Antarctic exploration proposal, but it came to nothing. So Mawson promulgated his own Australasian Antarctic Expedition. Unlike other Antarctic explorers, he was not aiming for the South Pole; he was most interested in the part of Antarctica lying directly below Australia, in what he called the Australian quadrant.

He initially envisaged three land parties, but modified this to two: the main party of 18, under Mawson, was based in Adélie Land, and the western party of eight, led by the experienced Antarctic expeditioner Frank Wild, was based in Queen Mary Land.

Mawson's ship was *Aurora*, an old Dundee whaler of 612 tonnes, captained by John King Davis, with a complement of four officers and 19 crew. Mawson also bought an aircraft; it crashed in October 1911 during a fund-raising display in Adelaide, but he took the wingless fuselage to Antarctica to use as a motor sledge. The expedition left Hobart on 2 December 1911, established a five-man radio station at Macquarie Island, and continued southward. The main party landed on the Antarctic mainland on 8 January 1912, and *Aurora* sailed west with Wild's party on 19 January.

"The home of the blizzard"

The main party had landed in what seemed an ideal natural harbor, which Mawson named Commonwealth Bay. He called the point where they erected their hut Cape Denison. Too late, he found that they were at the base of a funnel for winds that came down from the polar ice cap; he wrote: "The climate [was] little more than one continuous blizzard …" It was a trying winter, but the expeditioners kept meteorological observations, and erected aerials to send Morse messages to Macquarie Island, the first use of radio in Antarctica—though it was only toward the end that they received replies.

In August, they excavated and provisioned a comfortable cave in the ice of a slope five miles south of the hut. They named it Aladdin's Cave, and it was to prove a lifesaver. Because of spring blizzards, it was November before the five planned sledging expeditions set out. All left between 8 and 10 November, knowing that they must be back to meet *Aurora* by 15 January. One party was aiming for the South Magnetic Pole, and got to within 80 kilometers (50 miles) of their goal, at an

◀ BROADCASTING STATION
Ham Shack near the summit of Wireless Hill (112 meters/370 ft) marks the site that Mawson chose for his wireless station—the living hut for the five-man party was located where the base is today. Building materials and supplies for Ham Shack were carried up the steep hillside by an aerial cable, 243 meters (800 ft) long.

▲ THE LIE OF THE LAND
This photograph looking south from Wireless Hill that was taken during Mawson's expedition gives a good overview of the northern end of Macquarie Island. The main difference between now and then is the development of the year-round Australian base that is the only human habitation on this sub-Antarctic island.

Mawson described *Aurora* as "by no means young [but] still in good condition and capable of buffeting with the pack for many a year ... The hull was made of stout oak planks ... The bow ... was a mass of solid wood, armored with steel plates."

► "THE HOME OF THE BLIZZARD"
Landing stores and equipment at the head of the boat harbor, Cape Denison, were the first steps toward forming the main base station. In the distance men are sledging materials to where the hut is to be built.

altitude of 1,935 meters (5,900 ft), on 21 December before lack of time and food forced them to turn back. They returned to the hut on 11 January.

A second party planned to use the crippled aircraft to explore the hinterland of Adélie Land; once they even managed to drive it up a steep slope, but on the second day several pistons snapped, and they abandoned it. On 5 December they found a tiny black meteorite—the first ever discovered in Antarctica. They turned back on 26 December and after a difficult journey reached the base on 17 January.

The third and fourth parties traveled together for 75 kilometers (46 miles), and then the near-east group turned back to map the coast between the glacier named after the expedition's Swiss ski expert and

◄ ALADDIN'S CAVE
For Mawson the cave was "a truly magical world of glassy facets and scintillating crystals ... The purest ice for cooking could be immediately hacked from its walls ... Finally one neatly disposed of spare clothes by moistening the corner of each garment and pressing it against the wall for a few seconds, where it would freeze on and remain hanging until required."

◄ CHANGE OF PURPOSE
Long before this station was built, sealers had their huts at the southern end of the peninsula at the foot of Razorback Hill. According to Dr A.L. McLean, "everywhere [was] covered with the bones, bleached skeletons and putrid carcasses of sea elephants." Today, the isthmus is a peaceful place of scientific enquiry.

◄ EDGE OF A GLACIER
The Mertz Glacier is some 45 miles (72 kilometers) long and 20 miles (32 kilometers) wide, and where it reaches Cape Hurley, it extends into the water as the Mertz Ice Tongue seen here. These icy features were named by Mawson for Xavier Mertz who died on 7 January 1913 on the far-east sledge journey.

◄ FIELD WORK
At New Year the eastern coastal party set up camp at Penguin Point where Cecil Madigan described "a rock wall three hundred feet high formed the sea-face, jutting out from beneath the ice slopes of the buried land." Besides seals and Adélie penguins they found some new bird's eggs and mites living in the moss.

mountaineer, Xavier Mertz, and Cape Denison. In one small stretch they found an archipelago of 154 little islands. Meanwhile, the eastern coastal party continued eastward, making its difficult way across the rough Mertz Glacial Tongue, then sea ice, then the Ninnis Glacial Tongue, and then sea ice again as far as a towering coastal promontory of columnar lava that they named Horn Bluff—an expanse of spectacular "organ pipes," with outcrops of sandstone and coal. They started back on 21 December, but struck bad weather and soft snow, and passed several days with virtually nothing to eat before reaching a food depot. They were back in the comparative safety and warmth of the hut on 17 January.

A beleaguered party

The fifth party was overdue. Mawson had left on 10 November, with Mertz and the ex-Royal Fusilier Belgrave Ninnis. With three sledges and 16 dogs, they made good time at first, but disaster struck on 14 December, not long after they had abandoned one of the sledges. Mertz was leading on skis, Mawson was behind with the first sledge, and Ninnis was in the rear with the other. Mawson wrote: "My sledge crossed a crevasse obliquely and I called back to Ninnis … to watch it, then went on, not thinking to look back again …" But when he did look back, Ninnis had disappeared through a hole in the ice bridge that spanned the crevasse. Mawson and Mertz peered over the edge and saw two dogs and a tent and food bag caught on a ledge; there was no sign of Ninnis. Lacking a rope long enough to climb down, they called fruitlessly for four hours, read the burial service, and took stock of their own situation.

And their situation was desperate indeed, for Ninnis, in the supposedly safer rear position and with the best dog team, had been carrying most of the food, the main tent, and other vital supplies. "May God help us," wrote Mawson, as he contemplated their return journey. They started back on a more southerly route than their outward one to avoid the most dangerous crevasses, surviving by killing and eating the six remaining dogs; the paws took the longest to boil into an edible stew. It was 28 December when they dispatched Ginger, the last of the dogs.

▼ CLAIMED BY A CREVASSE
Belgrave Ninnis, a lieutenant of the Royal Fusiliers, was just 23 years old when he plunged to his death. Mawson wrote: "It was difficult to realize that Ninnis, who was a young giant in build, so jovial and so real but a few minutes before, should thus have vanished without even a sound."

◄ KILLED BY THE COLD
Dr Xavier Mertz was a 28-year-old lawyer from Berne, a ski jumping champion and an excellent mountaineer. His outer trousers and helmet were lost on Ninnis's sledge and, becoming wet and sick, he died soon afterwards. Mawson wrote: "No one could have done better … he had been a general favorite."

◄ RESTORED HUT AT CAPE DENISON

In the opinion of J. Gordon Hayes, Mawson's undertaking "judged by the magnitude both of its scale and of its achievements, was the greatest and most consummate expedition that ever sailed for Antarctica."

▼ CALLING ANTARCTICA HOME

When the main base hut was completed in February 1912, the Union Jack was hoisted as the men cheered. Mawson noted that soon after the hut was finished it "became so buried in packed snow that ever afterwards little beyond the roof was to be seen."

By 3 January 1913, Mertz was ravaged by dysentery and severe frostbite. Mawson told how Mertz tried to maintain his own courage at this time: "To convince himself he bit a considerable piece of the fleshy part off the end of one of [his fingers]." By 6 January Mertz was riding on the sledge; the following day he had several fits, and struggled to get out of his sleeping bag. Mawson woke a couple of hours later to find his companion "stiff in death." There have been insinuations that Mawson practiced cannibalism in these desperate straits, but he had enough food, and his actions in burying his comrade under blocks of snow and marking the position—and above all, his high principles—argue against this. He cut his sledge in half with a penknife to lighten his load, sewed a makeshift sail, and carried on.

The last lap

Mawson had recorded that the skin had peeled away from their groins, and on 11 January wrote: "My feet felt curiously lumpy and sore." The soles of his feet had come off, and had to be strapped on with bandages. But sudden death was more immediately threatening than frostbite. On 17 January Mawson fell through an ice bridge; suspended on a thin rope, and precariously anchored by

▲ MARKING TIME WITH MUSIC

Winter evenings passed slowly in the hut but books and a gramophone helped to entertain the expeditioners. Standing up: Mawson, Madigan, Ninnis, and Correll. Sitting round the table from left to right: Stillwell, Close, McLean, Hunter, Hannam, Hodgeman, Murphy, Laseron, Mertz, and Bage.

Roving camera

Frank Hurley, a Sydney suburban post-card photographer, had been accepted into the expedition after he inveigled his way into Mawson's overnight railway carriage, and went on to make a remarkable film of the expedition. He returned from Commonwealth Bay with most of the expeditioners, then sailed south again with *Aurora* to pick up Mawson, who said: "Hallo! Back again?" Two weeks after he returned to Australia with Mawson, he was filming a motoring trip into outback Australia when a black tracker arrived with a cabled invitation to join Shackleton's *Endurance* expedition. Two months later he sailed to Buenos Aires to meet the ship, and was soon trapped with it in the ice.

his half-sledge, Mawson thought: "So this is it … then I thought of the uneaten food on the sledge … A great effort brought a knot in the rope within my grasp, and after a moment's rest, I was able to draw myself up and reach another, and, at length, hauled my body on to the overhanging snow lip. Then when all appeared to be well … a further section of the lip gave way, precipitating me once more to the full length of the rope."

He contemplated giving up by releasing himself from the rope, but hauled himself to the surface, where he "cooked and ate dog meat enough to give me a regular orgy," and wove a rope ladder that he tied to himself and the sledge—and subsequently used on several occasions. By 29 January he was close to the Cape Denison hut. That day he found a cache of food and a message left

six hours earlier by three other members of the expedition, and learned that Amundsen had reached the South Pole, and that the rest of his own men were safe.

Mawson's problems were not over. The snow along his route turned to slippery ice, and he had discarded his crampons. He fell frequently, "until I expected to see my bones burst through my clothes," and made crampons from his wooden theodolite case and from nails and screws. He arrived at Aladdin's Cave on 1 February, only to be pinned down by a blizzard for a week. On 8 February he made his way to the Cape Denison hut, where he saw *Aurora* sailing into the distance, and "my hopes went down." Then three men waved from the harbor and ran to meet him: six men had stayed in case any of Mawson's party returned, and they wept when they heard of the deaths of Ninnis and Mertz.

A return in triumph

Aurora was immediately recalled by radio, but weather conditions made landing impossible, so she resumed her course to pick up Frank Wild's western party and returned to Hobart on 15 March 1913. Mawson believed that that storm may have saved his life, as he was gently nursed through the winter rather than having to withstand the rough Southern Ocean crossing. Sidney Jeffryes was the new radio operator, and his skill was such that they managed to communicate with Australia via Macquarie Island in the early part of the winter. However, Jeffryes suffered a mental breakdown and was able to work only intermittently in the second half of the year. *Aurora* returned to Commonwealth Bay on 13 December to collect Mawson's party, carried out marine research along the Antarctic coast for two months, and docked in Adelaide on 26 February 1914.

Mawson's scientific achievements had preceded him, and he was treated as a hero when he arrived home. On 31 March 1914 he married Paquita Delprat in Melbourne (J.K. Davis was best man); on 29 June 1914 (the day that Archduke Franz Ferdinand was assassinated in Sarejevo, triggering World War I) he was knighted at St James's Palace in London.

> **AT THE WHIM OF THE WIND**
Katabatic winds regularly whip Commonwealth Bay into a lather. Gusts there can also be very localized. Mawson, observing the behavior of the wind, noted: "Laseron one day was skinning at one end of a seal and remained in pefect calm, while McLean, at the other extremity, was on the edge of a furious vortex."

DOUGLAS MAWSON AND THE *AURORA* 1911–14

George V Land

Terre Adélie

150°E
140°E
70°S

Mertz Glacier

Mertz dies 7 Jan 1913

Ninnis Gl.

Ninnis dies 14 Dec 1912

Aladdin's Cave

food depot

main base
Cape Denison

Dixson I.

Mt Murchison

Horn Bluff

Commonwealth Bay

Mawson Pen.

Watt Bay

Fisher Bay

Cook Ice Shelf

Cape Freshfield

Ninnis Glacier Tongue

Mertz Glacier Tongue

Lauritzen Bay

Cape Hudson

140°E

Antarctic Circle

150°E

—— Far eastern party
—— Eastern coastal party
—— Near eastern party

0   200 kilometers
0   100 miles

The Western Party

While Mawson's party endured their various ordeals, *Aurora* had taken Frank Wild's party to the Shackleton Ice Shelf of Queen Mary Land, 2,400 kilometers (1,500 miles) to the west. Vast quantities of supplies had to be dragged by aerial cable to the summit of the ice shelf before the ship sailed away. Their hut on the ice was habitable by 28 February, but the radio masts they erected were blown down by the first blizzard. On clear days they could see the mainland, 27 kilometers (17 miles) to the south; they visited it in mid-March and left a cache of supplies.

Wild wrote: "The most bigoted teetotaler could not call us an intemperate party. On each Saturday night, one drink per man was served out, the popular toast being 'Sweethearts and Wives.' The only other convivial meetings of our small symposium were on the birthdays of each member, Midwinter's Day and King's Birthday."

In August, just before they planned on starting their spring skiing program, the hut nearly burned down while the acetylene generators were being charged. The first journey was to the east; six men and three dogs left on 22 August and returned on 15 September. In summer there were two expeditions: one to the eastern coast under Wild, and one to the western coast under Evan Jones. One man was left at the hut.

Wild's group found the going very heavy, particularly across the crevasses of the Denman Glacier. They climbed to the peak of Mount Barr-Smith on 22 December, at the end of their outward journey, and were back at the hut on 6 January 1913. Jones's party found vast colonies of penguins and other nesting birds on their way. They spent Christmas at a small, bare, extinct volcano that Drygalski had named Gaussberg when the Germans sledged there 10 years before; the German cairns were still visible on the summit. They were 346 kilometers (215 miles) from their hut, but were back there on 21 January.

When *Aurora* had not returned by the end of January, the party began to stockpile seal meat and blubber in case they were marooned for another winter. Saturday 23 February was the anniversary of the departure of *Aurora*, and in a howling blizzard they faced the possibility that the ship would not return. But *Aurora* arrived the following day, with most of the other expeditioners, and they were quickly boarded and taken back to Australia.

▲ VETERAN ANTARCTICAN Mawson thought Frank Wild "An excellent Petty Officer... and in some respects more than that ... He could not be excelled in intrepidity and had a full quota of sound horse sense. A very likeable fellow."

"By endurance we conquer."

When the news broke that Amundsen had reached the Pole, Shackleton wrote: "The discovery of the South Pole will not be an end to Antarctic exploration. The next work is a transcontinental journey from sea to sea, crossing the Pole." He soon turned his ambitions towards achieving that goal himself. The project was not without critics. He also faced the embarrassment of his brother Frank being jailed for fraud, and his own finances were always uncertain. But, with determination and Irish charm, Shackleton pulled it all together as the Imperial Transantarctic Expedition.

Shackleton's first plan was to take one ship, disembark at Vahsel Bay in the Weddell Sea, and travel 2,900 kilometers (1,800 miles) overland while the ship sailed around to meet them. If he was late and the ship had to depart, it would leave a small whaler behind and he could have a chance at another of his ambitions: an open boat journey across Antarctic waters. It was a foolhardy plan, and he was persuaded instead to use two ships, with a party on the Ross Sea establishing food depots for him on the Ice Shelf. Shackleton was now 40, but he optimistically predicted that he could complete the journey over completely unknown terrain in 100 days.

The ship for the Weddell Sea was *Polaris*, captained by New Zealander Frank Worsley and renamed *Endurance*, from Shackleton's family motto:

"By endurance we conquer." For the Ross Sea Party, he bought Mawson's *Aurora*. *Endurance* sailed from Plymouth on 8 August.

Shackleton made for the whaling settlement of Grytviken, about halfway along the north coast of South Georgia. He arrived in November 1914 to hear of the worst ice conditions ever reported in the Weddell Sea, but for Shackleton it was now or never. They sailed south on 5 December 1914 and were in solid pack ice by 11 December, with 1,600 kilometers (1,000 miles) to cover to Vahsel Bay. Shackleton had selected his polar party: surgeon Alexander Macklin; Australian photographer Frank Hurley; second-in-command Frank Wild; polar veteran Tom Crean; and George Marston, the artist from the *Nimrod* expedition.

The ice closes in

They saw land on 10 January 1915, and by 18 January they were only 130 kilometers (80 miles) from their goal. But next day the ice closed around *Endurance*; she was frozen in and began to drift gradually north with the pack ice. So began 10 months of entombment.

It was now that Shackleton's leadership qualities became apparent. The whole purpose of his expedition had been foiled within site of its goal. They might break free eventually—or the ship might be crushed in the ice. Meanwhile the other half of the expedition was laying supply depots on the Ross Ice Shelf that would never be used. Yet Shackleton betrayed no mental anguish; Macklin noted: "We could see our base, maddening, tantalizing, Shackleton at this time showed one of his sparks of real greatness. He did not rage at all ... he told us simply and calmly that we must winter in the Pack, explained its dangers and possibilities; never lost his optimism, and prepared for winter." They built small igloos—"dogloos"—on the ice for the huskies, and settled into a strict routine of shipboard life.

Slowly they drifted northward. About the time the sun reappeared, there were moments of terror as the pack ice formed pressure ridges and squeezed *Endurance* until her timbers groaned and bent. The pattern continued for three months, and the ship developed substantial leaks. Shackleton decided that they must be ready to abandon ship at any time. The following day they retreated to five tents on the ice. After their first night on the ice, Shackleton announced: "Ship and stores have gone—so now we'll go home."

Camping on the ice

After *Endurance* finally sank on 21 November, they could only wait for the ice where they were camped to drift north and break up. Towards Christmas they moved 16 kilometers (10 miles) from Ocean Camp to Patience Camp, planning to take to the three boats they had salvaged from *Endurance* in an attempt to reach land. Shackleton's leadership was critical: he had a knack of

ERNEST SHACKLETON
AND THE *ENDURANCE*
1914–16

spotting imminent psychological collapse and bolstering the sufferer's self-esteem.

The men lived on the ice for five months, surviving toward the end largely on fresh seal and penguin meat. Then the ice began to break up, so on 9 April 1916 they took to the boats and the numbingly cold, stormy seas. They had drifted with the ice for 3,220 kilometers (2,000 miles), and had now passed the tip of the Antarctic Peninsula. Their only course to steer was for Elephant Island, a desolate lump of rock; at least it was not an ice floe.

They were in the boats for seven days and nights, sometimes camping on ice floes, before landing at Elephant Island's Cape Valentine. It was just in time. Shackleton elected the youngest member of the expedition, Perce Blackborrow, to have the honor of being first ashore. He half pushed him off the boat, but the man was unable to walk because his feet were frostbitten, while Hurley bounded about recording the scene with his pocket camera. Shackleton recorded that some of the men were reeling about the beach. Others were so grateful to be on land after almost 500 days that they lay down on the shingles and poured stones over themselves.

They could not stay where they were, as they could be washed away from this exposed shore. They found a more sheltered spot nine miles down the north coast, with a gravelly beach; it was still desolate, but there was plenty of wildlife, which they would need. The night they moved there and named it Cape Wild, a gale shredded their remaining tents. Morale was at its lowest.

Shackleton stated that he and five of the fittest men would sail to South Georgia for help in the *James Caird*—more than 1,290 kilometers (800 miles) in an open boat, through some of the wildest seas in the world. They set out on 24 April, with Worsley as captain and navigator; Shackleton, Crean, and McCarthy went too, with Harry McNeish and John Vincent.

Unable to take more than four sun sightings over a fortnight, and with the setting of his chronometer uncertain, Worsley steered by dead reckoning and gut instinct, knowing that if he was wrong they would sail past South Georgia into the South Atlantic, and the men on Elephant Island would never be found. Of the 17 days of their journey, 10 were in full gales. One night they awoke while hove to in a gale to find the boat foundering and thick ice encasing every sodden inch of wood and canvas. They had to crawl out on the glassy decking and hack the ice away with an axe.

When they sighted South Georgia on 8 May, after 15 days at sea, they were on the eastern side of the island. They landed at King Haakon Bay on the evening of 10 May after several near-disasters along the coast. The Norwegian whaling stations were on the other side of the island, over a wall of high mountains. Only three were fit to attempt the crossing: Shackleton, Worsley, and Crean. Their fourth attempt to find a pass through the mountains took them over just as daylight was failing.

Shackleton knew if they stayed at high elevation overnight they would freeze to death. Beneath them lay a long, steep precipice of snow. "It's a devil of a risk but we've got to take it. We'll slide," he said. Coiling their climbing rope beneath them, they sat one behind the other, and pushed off into the darkness.

"We seemed to shoot into space," wrote Worsley. "For a moment my hair fairly stood on end. Then quite suddenly I felt a glow, and knew that I was grinning! I was actually enjoying it … I yelled with excitement and found that Shackleton and Crean were yelling too." They descended 460 meters (1,500 feet) in a matter of seconds. They stumbled on, falling and slipping, and barely conscious. At dawn Shackleton recognized a rock formation at Stromness Bay, and at 6.30 am heard the welcome sound of the steam whistle blast to wake up the Stromness workers. They were saved.

A ship was sent around to the east coast to collect the other three crew of *James Caird* and the little boat itself. But it was to be another five months before ice conditions made rescue of the men marooned on Elephant Island possible.

Amazingly, Shackleton had not lost a single man during the two-year ordeal. However, his task was not over: the Ross Sea Party had to be rescued, and money raised for the *Aurora* to go south again under the command of John King Davis. Shackleton sailed directly from South America to New Zealand to join the expedition. Davis found Shackleton "somewhat changed," and allowed him to sign on only as a supernumerary, but recorded that at sea Shackleton's "old greatness of spirit shone out."

▲ SOUTH GEORGIA AHOY!
George Marston painted *James Caird* arriving at the rugged south coast of South Georgia from material supplied by those who were there. Shackleton recorded that the first view "was a glad moment. Thirst-ridden, chilled and weak as we were, happiness irradiated us. The job was nearly done."

▼ NO WAY TO CROSS
The night before he left to cross the island with Crean and Worsley, Shackleton was unable to sleep: "My mind was busy with the task of the following day … No man had ever penetrated a mile from the coast of South Georgia at any point, and the whalers I knew regarded the country as inaccessible."

The Ross Sea ordeal

The other part of the expedition, the Ross Sea Party on *Aurora*, experienced as many perils as those on *Endurance*; indeed, three men died.

Aurora was to moor off Ross Island for the winter as a base for a sledging party to lay food depots across the Ross Ice Shelf and onto the Beardmore Glacier for Shackleton's homeward journey. The ship left Hobart on Christmas Day 1914 and arrived at Ross Island on 9 January 1915, captained by Aeneas Mackintosh. Also on board was an experienced polar veteran, Ernest Joyce, but Shackleton was vague about which of the two was in command, which led to friction and disaster.

Mackintosh decided to use Scott's Cape Evans hut. On 24 January the first depot-laying expedition set out. Mackintosh was eager to get moving in case Shackleton arrived at the end of this first summer, and—against Joyce's advice—he pushed the dogs too hard; all but five died, and much of the task remained uncompleted.

Aurora vanishes

Aurora was to be a supply and accommodation base until spring, when the laying of supply depots would begin, and by mid-March she was secured at Cape Evans by her two anchors and seven steel hawsers. The 10 men who would lay the depots were ashore with minimal equipment. But on the night of 6 May there was a violent storm, and by 3.00 am, the ship had vanished— its mooring of ice had blown away, leaving only bent anchors and snapped hawsers. It had happened before, and they expected *Aurora* to return, but the vessel was as firmly gripped in ice as *Endurance* on the other side of the continent; almost a year would pass before *Aurora* broke free and limped back to New Zealand.

After another blizzard, they knew that they were marooned for the winter. Mackintosh's first depot-laying expedition was still struggling back, and the four at Cape Evans frantically searched to see what Scott had left behind. They found tins of jam, flour, and oatmeal, and some pemmican, but no soap, tobacco, or medical supplies, and their only clothing was what they were wearing and some spare underwear. They killed seals for meat and fuel—and kept taking scientific observations.

By 2 June, the sea ice was firm enough for Mackintosh's party to get back to Cape Evans, where they learned of the loss of *Aurora*. Rescue might be two years away, but they still had to lay supply depots for Shackleton in the spring or he would certainly die on the ice shelf. Supplies left by Scott, including kerosene and two Primus stoves, were a life-saving discovery, and they also found cake, chocolate, sleeping bags, socks, underwear, and a tent that they cut up and sewed into sledging clothes.

After taking an inventory, they planned to carry 1,800 kilograms (4,000 lb) of supplies onto the Ice Shelf and establish the furthest depot at Mount Hope, at the foot of the Beardmore Glacier, at 83°40′S. Joyce was deferring to Mackintosh, but relations were strained. In mid-August Mackintosh and chief scientist Fred Stevens reached Shackleton's hut at Cape Royds and found cigars, tobacco, food, and soap.

During September they hauled supplies across to Hut Point, the starting point for sledging south. Joyce wrote: "Most of us wore the canvas trousers made from Scott's old tent, and they froze on us like boards." In late October they found a sledge left as a marker by Cherry-Garrard, a note for Scott, and six boxes of dog biscuits impregnated with cod liver oil; Joyce wrote: "At last we have struck gold in the Antarctic."

▲ DEEP FREEZE
The desiccated remains of one of the Ross Sea Party's dogs still lies in its collar by the Cape Evans hut. The work of the expedition was made infinitely harder by Captain Mackintosh pushing the dogs too hard too soon, so that most died before the main task of laying supplies.

The first death

On 3 January 1916 one Primus was burning its own metal, and the party that was depending on it had to turn back. The men were failing, too: by 82°S Mackintosh was weakening and Arnold Patrick Spencer-Smith, the chaplain, had scurvy. They

◄ UNCERTAIN FATE
When the survivors were reunited at Cape Evans, those who crossed from Hut Point were sad but not surprised that Mackintosh and Hayward had not made it. With considerable under-statement, Dick Richards later recalled that "From July 1916 until January 1917, when the rescue ship arrived, it was a bit of a struggle to get by."

left him in a tent and pushed on to establish the southernmost supply depot at Mount Hope on 26 January. When they returned, Spencer-Smith could not walk; they carried him on the sledge until he died on 8 March and then they buried him in the ice.

By now they were all suffering from scurvy, and when a blizzard stopped them close to where Scott and his companions had perished, they seemed set for the same fate, but they reached Hut Point on 11 March. They had been away for over six months, "without change of clothing or a bath." The four who had turned back earlier had reached Cape Evans, and now the survivors at Hut Point could eat seal meat and recover their strength. Joyce calculated that they must wait four months for the surface to be stable enough to cross to Cape Evans, but by 8 May after a blizzard had cleared much of the sea ice, Mackintosh announced that he and Hayward would walk across the 19 kilometers (12 miles) that remained. It was a suicidal decision: a blizzard arose, and Mackintosh and Hayward were never seen again. On the night of 15 July, the remaining three, Dick Richards, Ernest Joyce, and Ernest Wild, crossed to Cape Evans and rejoined the other four. The seven men then had to endure a long wait for rescue.

Meanwhile, *Aurora* and her crew had drifted 1,130 kilometers (700 miles) in the ice. North of the Antarctic Circle, as summer approached, they hoped that the ice would open up and release the ship, but the looser ice did even more damage; when *Aurora* was finally freed in March 1916, she was leaking and relied on a makeshift rudder to cover the 5,220 kilometers (2,000 miles) to New Zealand. On 3 April 1916 she was taken under tow to Port Chalmers, near Dunedin.

Shackleton returns

Rescue came to the Ross Sea on 10 January 1917. Dick Richards was outside looking for seals when he saw a ship's smoke. He walked calmly back into the hut and casually said: "There's a ship out there." When they saw that the ship was *Aurora*, the men shouted for joy. The ultimate miracle was when they realized that one of the men approaching them was Shackleton—coming from the north, not lost in the south as they had thought.

The Ross Sea Party had been on the ice for 199 days, without adequate clothing or support, and burdened with food for the failed polar party—indeed, an incredible achievement.

The final quest

In 1921 Ernest Shackleton made his final visit to the Antarctic region, in the *Quest*. At Grytviken in South Georgia he suffered a massive heart attack, and died at 2.50 am on 5 January 1922. He was 47 years old.

After holding a memorial service and erecting a cairn to his memory, *Quest* continued its voyage, but achieved little.   DM

▲ FINAL RESTING PLACE Shackleton was buried in the whalers' cemetery at Grytviken in South Georgia (symbolically, his was the only grave with its head facing south rather than east). When *Quest* returned in April, his seven comrades from *Endurance* paid their respects at the graveside. Frank Wild, his trusted lieutenant, is second from the left.

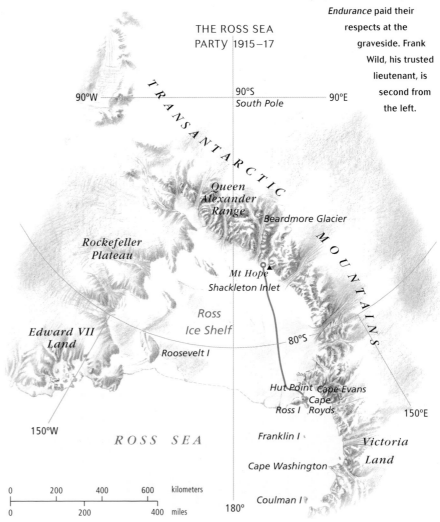

THE ROSS SEA
PARTY 1915–17

Modern Explorers

1921–59

1934–37: The Moth aircraft used by Rymill could fly with either skis or floats, and was small enough to be towed by *Stella*, the expedition's motor launch, as seen here. This versatility was enhanced by the use of dog and motor sledges.

1928: Sir Hubert Wilkins was the first person to fly an airplane in Antarctica. He was also the first to successfully fly across the Arctic Ocean, for which he was knighted in 1928. After his Antarctic exploits he was invited to join the airship *Graf Zeppelin* on the first round the world flight in 1929.

1929–31: The BANZARE scientists went ashore for 10 days on Heard Island where they used a small hexagonal hut at Atlas Cove, built by Norwegians as a refuge for shipwrecked whalers. After baracading the door against marauding elephant seals, Frank Hurley praised his temporary home as "warm, dry, rent free, and no taxes".

1926: When Grytviken was a working station, whales were hauled onto the open space between the jetties and stripped of their blubber—flensed. The blubber was fed into huge pressure cookers in the building with chimneys (center). The oil was pumped into the tanks behind. Meat, bones, and guts were turned into fertilizer and animal feed.

1926: Scientific research on whales and whaling led to the launch of the Discovery Investigations. However, by the time the *Discovery* voyages ceased in 1951, factory ships and steam whalers had drastically reduced the world's whale populations.

1938–39: Wilkins claimed the Vestfold Hills area for Australia in the summer of 1938–39. This cairn is known as Walkabout Rocks because he protected his handwritten proclamation by wrapping it in a copy of the Australian magazine *Walkabout*.

On 5 January, Ernest Shackleton suffers a massive heart attack and dies at 47 years of age. At the time, he is on board the *Quest* in South Georgia on his last Antarctic expedition.

On 16 November, Hubert Wilkins and Carl Eilson make the first powered flight in Antarctic skies. A month later, they make the historic flight of over 2,100 kilometers (1,300 miles) that proves the feasibility of exploring the south polar regions by air.

Douglas Mawson leads two voyages of the British, Australian and New Zealand Antarctic Research Expedition (BANZARE) on board Scott's old ship, *Discovery*. The expeditioners carry out survey work, scientific research, and aerial mapping. They raise the British flag at Commonwealth Bay, and at several other locations.

After two failed attempts and several accidents, Lincoln Ellsworth finally achieves his dream of making the longest transantarctic flight thus far, in his Northrop Gamma monoplane *Polar Star*.

While acting as advisor on Lincoln Ellsworth's final expedition in Antarctica, Hubert Wilkins claims the Vestfold Hills area (now Davis station) for Australia.

| 1921 | 1922 | 1926 | 1928 | 1929 | 1929–31 | 1934–37 | 1935 | 1936 | 1938–39 |

After the British Imperial Expeditioners of 1920–22 fail to cross the Antarctic Peninsula, two members volunteer to stay for the winter and carry out scientific research. Thomas Bagshawe and Lieutenant Michael Lester keep a two-hourly meteorological log, an ice log, a natural history log, and, for a month, an hour's tidal log.

The first *Discovery* Expedition arrives in South Georgia. These expeditions, named after Robert Scott's ship *Discovery*, were instigated by the Whaling Committee with the aim of "conducting research into the economic resources of the Antarctic with the particular object of providing a scientific foundation for the whaling industry".

Richard Byrd as navigator, Bernt Balchen as pilot, and two other men become the first to fly over the South Pole when they take off in a Ford Trimotor on 29 November. With fuel so important a consideration, they spend just nine minutes at the Pole.

John Rymill, an Australian farmer and grazier with a lifelong passion for polar regions, leads the British Graham Land Expedition (BGLE) on board the *Penola*. It is the first expedition to combine the traditional use of dogs and skis with modern tractors, motor boats, and aircraft.

On its fourth commission, *Discovery II* is diverted from its scientific work as part of the Discovery Expeditions to rescue Lincoln Ellsworth and Hollick Kenyon from the Bay of Whales when their plane crashes nearby.

Background image: An aerial view of the ice-locked coastline and Transantarctic Mountains north of Terra Nova Bay on the western shoreline of the Ross Sea. The use of aircraft drastically changed the nature of Antarctic exploration, and allowed vast areas of the continent to be mapped.

1935: Ellsworth's Northrop Gamma monoplane, *Polar Star*, finally fulfilled its destiny to make the first transantarctic flight. Despite running out of fuel only 40 kilometers (25 miles) short of its destination, it was recovered by the *Wyatt Earp* and is now on display at the National Air and Space Museum in Washington DC.

1957–58: Sir Edmund Hillary, New Zealand conqueror of Everest (left) and Dr Vivian Fuchs, Director of the British Antarctic Survey, were a formidable team to lead the first successful crossing of the Antarctic Continent.

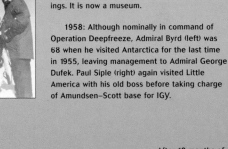

1935: Ellsworth (left) and his co-pilot Norwegian Bernt Balchen in the cockpit of *Polar Star*; in January 1934 they planned to fly across Antarctica from the Ross Sea to the Weddell Sea and back. But the plane was badly damaged when the ice shelf on which it had been unloaded from the ship broke up.

1946–47: Supplies are unloaded onto the sea ice in the Bay of Whales (USS *Yancey* in foreground). All equipment had to be taken by tractor-drawn 10-ton sledges to the site of Little America IV, 2 kilometers (over 1 mile) away. After the three-month operation the base was abandoned, never to be used—or even seen—again.

1957–58: Scott Base, established by Sir Edmund Hillary on the southern tip of Ross Island, is about one-tenth the size of its mighty US neighbor, McMurdo. The hut in which Hillary and his companion spent the winter of 1957 can be seen on the far left in front of the main buildings. It is now a museum.

1958: Although nominally in command of Operation Deepfreeze, Admiral Byrd (left) was 68 when he visited Antarctica for the last time in 1955, leaving management to Admiral George Dufek. Paul Siple (right) again visited Little America with his old boss before taking charge of Amundsen–Scott base for IGY.

Admiral Richard Byrd leads Operation Highjump, an ambitious venture organized by the United States Navy. The first primarily military Antarctic operation involves 13 ships, 23 aircraft, and over 4,700 personnel. For all its massive scale, however, its success is limited: only about a quarter of its initial objective is photo-mapped.

Paul-Emile Victor leads the first French Polar Expedition to Antarctica. Having sailed from Hobart in December 1949, the ship *Charcot* arrives at Cap de la Découverte in Terre Adélie after negotiating the pack ice for a few weeks. The landing party spends the winter at Port Martin.

As part of ANARE, Mawson station is opened at Horseshoe Harbor, in Mac.Robertson Land, on 13 February.

The IGY also sees the fulfilment of Ernest Shackleton's dream of 40 years earlier when a joint Commonwealth team (the Commonwealth Transantarctic Expedition) makes the greatest traverse of all—crossing the Antarctic Continent from coast to coast.

After 18 months of negotiations, the Antarctic Treaty is signed in December 1959 by the 12 nations that had participated in the IGY. The Treaty bans military activity, guarantees free access for scientific research, and holds in abeyance all territorial claims in Antarctica.

1938–39   1946–47   1947   1950   1951   1954   1957   1957–58   1958   1959

The Schwabenland Expedition, the shortest Antarctic expedition ever undertaken, is mounted by Air Chief Marshall Hermann Göring in a bid to claim Antarctic territory for Germany. It is the first attempt by any nation at large-scale aerial mapping of Antarctica. Over just seven days, two aircraft make 16 flights and survey 600,000 square kilometers (232,000 sq. miles).

In order to safeguard Australia's claim to 42 percent of the Southern Continent, the Australian National Antarctic Research Expeditions (ANARE) are launched.

The Discovery Expeditions cease. Their work laid the foundation for scientific whale conservation and led to the establishment of the International Whaling Convention in 1936, and eventually resulted in effective regulation of the whaling industry.

The International Geophysical Year (IGY) begins on 1 July, and runs for 18 months. Its scientific program focuses on Antarctica and outer space. Scientists from over 50 countries work together in an ambitious and successful program of cooperation.

The United States meets its IGY objectives with the conclusion of a four-year program called Operation Deepfreeze. This major undertaking yields important scientific data about Antarctica.

Antarctic flight

Exploration by air greatly expanded our knowledge of Antarctica. This advance was led by two intrepid polar explorers. Possibly the greatest adventurer and explorer Australia has ever produced, Hubert Wilkins was knighted in 1928 in recognition of his pioneering flight over polar ice from Alaska to Spitsbergen, in Norway. He had a distinguished photographic and flying career during World War I. American flying ace Richard Byrd claimed to have been the first to fly over the North Pole, on 9 May 1926, and in 1927 he was pipped at the post by Charles Lindbergh in the famous race to fly solo and nonstop across the Atlantic from New York to Paris.

Hubert Wilkins

With American aviator Carl Eilson, Hubert Wilkins flew with great distinction in the Arctic from 1925 to 1928. The fame that this attracted secured financial backing from American newspaper magnate William Randolf Hearst. In return for the news and radio rights, Hearst raised the money for an Antarctic expedition in the southern summer of 1928, after the Australian government declined to finance Wilkins's dream of exploring and photographing the south polar regions and perhaps reaching the South Pole.

The first flight: 1928

Early in November 1928, Wilkins's small expedition of just five men, with Eilson as chief pilot, reached its proposed base in the relative shelter of Deception Island, the volcanic caldera 80 kilometers (50 miles) northwest of the Antarctic Peninsula. *Hektoria*, the Norwegian whaling ship that took them there, carried two single-engined Lockheed Vega high-winged monoplanes, one of which Wilkins and Eilson had used to good effect in the Arctic. Twelve days later, on 16 November, the two aviators made the first powered flight in Antarctic skies. They took off from a very rough and less-than-straight airstrip scraped out of the black sand beach. The first flight over Antarctica lasted a mere 20 minutes; after a few brief circuits of the island, and with the weather deteriorating, the plane was back on the ground. But the exploration of Antarctica by air had begun.

Ten days later both aircraft flew, again locally over the island. On landing, however, one of them skidded from the ice into the water, where it lay suspended only by its wings, to be salvaged with great difficulty. Finally, on 20 December, after persistent frustrations with the weather, Eilson and Wilkins were able to make the historic flight of over 2,100 kilometers (1,300 miles) that was to prove the feasibility of exploring the south polar regions by air. They flew for 11 hours from Deception and back, across to the mainland of the Antarctic Peninsula, and then south to the Antarctic Circle and beyond, over much territory that had never before been seen,

let alone explored, Wilkins recording their journey in photographs and sketches. Some of his observations were inaccurate, but the value of the airplane in Antarctic exploration had been demonstrated beyond doubt.

Richard Byrd

By the late 1920s, Byrd had one overriding ambition: to be the first to fly over the South Pole. It was a more difficult proposition than the North Pole, but Byrd was backed by such great American financial names as Edsel Ford, who provided the Ford Trimotor aircraft that was to be the backbone of Byrd's operation, and the millionaire philanthropist John D. Rockefeller, who had extensive media interests. Byrd established a camp, which he called Little America, on the Ross Ice Shelf, a little way inland from the Bay of Whales, at the end of December 1928. The expedition of more than 80 men carried sophisticated radio equipment for maintaining contact with the outside world—mainly America—to recompense his media backers.

First over the South Pole: 1928–30

Byrd's first Antarctic flight took place on 15 January 1929. For the next month, the expedition concentrated on the hitherto unexplored lands east of Little America, using the Ford Trimotor and the other two expedition aircraft, a single-engined Fokker Universal, and a Fairchild. By March, however, flying conditions had become virtually impossible; the Fokker was destroyed on the ground in a blizzard, and the remaining two aircraft were deliberately buried in large pits to protect them for the winter. About half the crew wintered underground, and it was not until early October that a land-based geological party was able to embark on an extended expedition to the Queen Maud Mountains, and final plans for the polar flight could be finalized. Byrd was doubtless spurred on by the news that Wilkins intended to return to Antarctica.

The Ford Trimotor lacked the range to fly to the Pole and back nonstop, so on 19 November, mounted on skis, it first flew to the bottom of the Axel Heiberg Glacier to establish a fuel dump and thereby reduce fuel dependency for the polar flight by 700 kilometers (435 miles).

The attempt on the Pole began on the afternoon of 28 November with a crew of four, including Bernt Balchen as pilot and Byrd as navigator. The main barrier was the inland plateau, and the greatest drama of the flight unfolded as Balchen tried to fly up the Liv glacier into the skies above the plateau. Severe turbulence hammered the aircraft, and it was unable to climb. They faced an unenviable choice: turn back, or jettison weight in the form of either vitally needed fuel or food. It was the food that went, and the aircraft finally gained enough height—just—to clear the top of the glacier.

The rest of the flight was largely uneventful, and they flew over the South Pole at 1.14 am on 29 November, nine hours and 56 minutes after departure.

Back to the Pole: 1933–35

Byrd returned to the acclamation of a proud and grateful nation. The fame and fortune he craved was his for the asking, in the form of honors, adulation, and further financial backing. So he mounted another expedition. Rather than making record-breaking flights, the aim of this expedition was to explore and map the great uncharted *terra nullius* that extended far to the east. Scientific research was another major goal, particularly study of the meteorology and geology of Antarctica. From January 1934 until February 1935 (except for the very significant over-wintering time, when no flying was possible), well over one million square kilometers (386,000 sq. miles) were photographed and surveyed from the air, including vast tracts of Antarctic lands previously unexplored, indeed unseen. In the process, the autogiro was wrecked and the Fokker crashed in flying accidents, but no lives were lost.

Government takes the reins: 1939–41

Byrd's 1933–35 undertaking was among the last of the large, privately funded and operated Antarctic expeditions: thereafter governments became involved. German interest in Antarctica worried the United States government, and in late 1939 Byrd went south again, this time as head of a government-funded expedition. Once more, aircraft played a leading role; the expedition included a fleet of two Curtiss Condors, a Beechcraft monoplane, and a twin-engined, all-metal Barley-Grow floatplane. By now, Antarctic ventures had become almost routine, and the showmanship of earlier expeditions had faded away. Byrd's third expedition conducted extensive survey work and scientific research and evaluation from two bases until March 1941, when the expedition came to an end. **PL**

⌃ MEMORIES AND RELICS

This hangar and derelict Beaver are relics of the British Antarctic Survey base that operated on Deception Island from 1943 until 1967, when a volcanic eruption covered the base in cinders, forcing a hurried evacuation. The black sand beach was used as an airstrip by both the BAS Beavers and Wilkins's Lockheed Vegas in 1928.

BANZARE

> **TEAM SPIRIT**
> The expedition carried a scientific staff of 12, seen here with other members of the crew. The photographer Frank Hurley, who had served with both Mawson on the AAE and Shackleton on the *Endurance* expedition, is in the front row (with scarf). This image of Mawson in a balaclava (behind Hurley) featured on Australia's first $100 note.

The 1920s saw a surge of interest in Antarctica by several nations. Norway was eager to extend its whaling grounds from the South Atlantic to the Ross Sea and East Antarctica. France wanted to protect its interests in Adélie Land, based on d'Urville's voyage of 1840, as well as in Iles Kerguelen and other sub-Antarctic islands. In response, in 1923 Britain established a territorial claim to the Ross Dependency, both to control and to profit from the expansion of whaling, and at the 1926 Imperial Conference articulated a policy to "paint Antarctica red"—a reference to the practice at the time of mapping the British Empire in red. In particular, Britain had set its sights on the so-called Australian sector, which had been explored by Douglas Mawson's 1911–14 Australasian Antarctic Expedition (AAE). Mawson himself was keen to extend his discoveries, emphasizing that "over half the circumference of the globe remains to be charted in high southern latitudes," and that, in particular, "the great Antarctic region lying to the southwards

… is a heritage for Australians … nearer to the Commonwealth than the distance between the east and west coasts of our own continent."

A hard-working ship
Mawson's persistence led to an agreement between the British and Australian governments to send an expedition to chart the coast of Antarctica from 160°E to 85°E. This was BANZARE—the British, Australian and New Zealand Antarctic Research Expedition. The Falkland Islands agreed to lend Scott's old ship, *Discovery*, which was being used for the Discovery Investigations, for two seasons, from 1929 to 1931. Mawson was expedition leader, and the captain for the first voyage was J. K. Davis, Mawson's skipper from the AAE. The expedition left London in August 1929.

As well as survey work, the plans included a full program of biological and hydrographic research, so there were 12 scientists on board, as well as the ship's crew of 17, plus a Gipsy Moth biplane for aerial reconnaissance and mapping. The ship was not just overcrowded; Davis was convinced that "a more unsuitable ship for oceanographical work than *Discovery* could not have been built." She was very seaworthy, but she rolled so badly that it was like "a sea going rodeo," and was slow, underpowered, and lacking in coal-carrying capacity.

BANZARE sailed from Cape Town in October, first to Iles Kerguelen and to Heard Island, then southward to Enderby Land—which had not yet been claimed—to forestall the Norwegian expedition, under Riiser Larsen, that was already in the area. Mawson sighted land several times and made three flights over it, but he was able

⋀ LOST OPPORTUNITY
Sir Douglas Mawson, who was knighted after his Australasian Antarctic Expedition of 1911–14, was 47 when he returned as leader of BANZARE. Although committed to its program of scientific studies, the expedition offered Mawson little of the overland exploration at which he excelled, and he was disappointed with its results.

to land only once, on Possession Island (53°37′E) on 13 January 1930. Next day, at 49°E heading west, he met Larsen in *Norvegia*, heading east; by mutual agreement, and in accord with their governments' respective policies, the two expeditions turned around, marking 45°E as the boundary between Norway's claim to the west and Britain's claim to the east. In mid-January, with coal running out, *Discovery* turned homeward, arriving in Adelaide two months later.

Staking a national claim
Discovery sailed south again from Hobart in November, led again by Mawson, but with K. N. Mackenzie as captain and a similar complement of scientists and surveyors. The route led south to Macquarie Island, where they found the derelict AAE huts and wireless masts, and then to the Ross Sea, where *Discovery* refueled from the huge whaling factory ship *Sir James Clark Ross*—a striking contrast between old and new. Following the coast westward, they encountered heavy pack ice and could not land until they reached Commonwealth Bay, the site of the AAE's main base in 1911–14, where Mawson found his old hut filled with ice. A Proclamation claiming the region for Britain was read and the expedition pressed on, sailing over "land" reported by Charles Wilkes in the 1840s; Mawson reasoned that it must have been an ice tongue. Further west, they landed and raised the flag at Scullin Monolith and Cape Bruce; they also sighted the Casey, David, and Masson ranges of Mac.Robertson Land, and, with the help of the Gipsy Moth, discovered the large indentation of Prydz/Mackenzie Bay.

Bad weather and heavy ice hampered both of the BANZARE voyages, and Mawson was disappointed with the results; he was unable to undertake as much overland exploration as he would have liked, and had to rely on aerial reconnaissance. Even so, the expedition gathered a wealth of information about a previously unknown part of Antarctica, and resulted in the 1933 affirmation of Britain's claim to the Southern Continent between 160°E and 45°E; later, in 1936, this was to become the Australian Antarctic Territory.

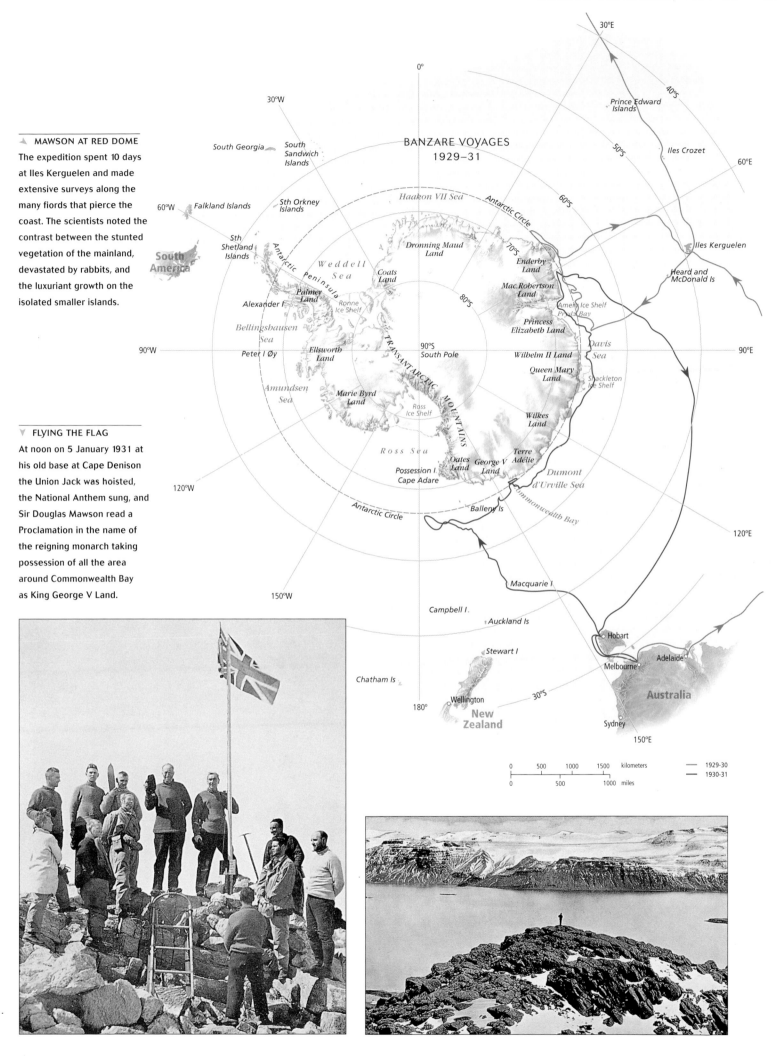

△ MAWSON AT RED DOME
The expedition spent 10 days at Iles Kerguelen and made extensive surveys along the many fiords that pierce the coast. The scientists noted the contrast between the stunted vegetation of the mainland, devastated by rabbits, and the luxuriant growth on the isolated smaller islands.

▽ FLYING THE FLAG
At noon on 5 January 1931 at his old base at Cape Denison the Union Jack was hoisted, the National Anthem sung, and Sir Douglas Mawson read a Proclamation in the name of the reigning monarch taking possession of all the area around Commonwealth Bay as King George V Land.

BANZARE VOYAGES
1929–31

30°E
40°S
Prince Edward Islands
50°S
Iles Crozet
60°E

30°W
0°
South Georgia
South Sandwich Islands
Haakon VII Sea
Antarctic Circle
60°S
70°S
Iles Kerguelen
Heard and McDonald Is

60°W
Falkland Islands
Sth Orkney Islands
Dronning Maud Land
Enderby Land
Mac.Robertson Land
Amery Ice Shelf
Prydz Bay

Sth Shetland Islands
South America
Antarctic Peninsula
Weddell Sea
Coats Land
Ronne Ice Shelf
Palmer Land
Alexander I
Princess Elizabeth Land
Davis Sea

90°W
Bellingshausen Sea
Peter I Øy
Ellsworth Land
80°S
90°S
South Pole
TRANSANTARCTIC MOUNTAINS
Wilhelm II Land
Queen Mary Land
Shackleton Ice Shelf
90°E

Amundsen Sea
Marie Byrd Land
Ross Ice Shelf
Wilkes Land

120°W
Ross Sea
Oates Land
George V Land
Terre Adélie
Dumont d'Urville Sea
120°E

Possession I.
Cape Adare
Commonwealth Bay
Balleny Is

150°W
Antarctic Circle
Macquarie I

Campbell I.
Auckland Is
Hobart
Adelaide
Melbourne
Australia

Chatham Is
Stewart I
180°
Wellington
New Zealand
30°S
Sydney
150°E

| 0 | 500 | 1000 | 1500 | kilometers |
| 0 | 500 | 1000 | miles |

——— 1929-30
——— 1930-31

ANARE

During the "cold war" that followed World War II, both the United States and the Union of Soviet Socialist Republics looked to Antarctica as a possible sphere of economic advantage and influence. This put pressure on Australia to safeguard its claim to 42 percent of the Southern Continent. Douglas Mawson was instrumental in persuading the Australian government to act decisively, and he joined other polar veterans, including J. K. Davis, on the Executive Planning Committee charged with implementing a policy of "effective occupation" of Australia's Antarctic sector. As a result of his efforts, the Australian National Antarctic Research Expeditions (ANARE) were launched in 1947. Stuart Campbell, another Antarctic veteran, was chosen to lead the first ANARE, which set up bases on Heard Island in December 1947 and Macquarie Island in March 1948, each with a staff of fourteen. The only ship available was a war-surplus HMAS LST 3501 (landing ship, tanks), which "quivered like a springboard … the decks bend and ripple like a caterpillar in progress" in the swells of the Southern Ocean. A simultaneous attempt to reach the Antarctic Continent in *Wyatt Earp*,

▼ **DISTINGUISHED SERVICE**
Nella Dan was the longest serving, and best loved, of the *Dan* ships that provided access to the continent for ANARE for over 30 years. Her stranding and subsequent scuttling at Macquarie Island in 1986 marked the end of an era. In 1990 she was replaced by a purpose-built research and resupply icebreaker *Aurora Australis*.

➤ **CELESTIAL DISPLAY**
The Antarctic atmosphere produces spectacular effects like this solar pillar caused by refraction of light by ice crystals, photographed near Davis station. Davis, in the ice-free oasis of the Vestfold Hills, is known as the 'Riviera of the South' for its warm sunny summers and comparatively mild winters.

the old wooden sealing ship that Lincoln Ellsworth had used in 1934–35, failed due to heavy ice and mechanical problems. It was not until 1954, with the charter of the Danish ice-strengthened cargo vessel *Kista Dan*—the first in a long line of *Dan* ships that serviced ANARE until 1986, when *Nella Dan* was wrecked at Macquarie Island—that access to the Southern Continent could be guaranteed and a permanent base established.

Horseshoe Harbor, in Mac.Robertson Land, was selected as an ideal location, and there Mawson station opened on 13 February 1954. The Heard Island station was closed in 1955, and much equipment was transferred to the new base, including dog teams and sledges. These became the backbone of an extensive program of exploration and mapping, which reached south to the Prince Charles Mountains and east and west along the coast of Enderby Land and Mac.Robertson Land. Dog sledges were supplemented by tractors, and from 1956 by RAAF Beaver and Auster aircraft. The Lambert Glacier—the world's largest—was discovered, and enormous tracts of land, including 2,000 kilometers (1,240 miles) of coastline were photographed.

Australia's Antarctic role

Australia has had a continuing commitment to the scientific study of Antarctica, and this was reflected in her major participation in the International Geophysical Year (IGY) of 1957–58, and in her later role in negotiating the Antarctic Treaty and as founding signatory in 1959. During the IGY a second Australian station, Davis, was built in the Vestfold Hills, and the program of Antarctic meteorological, magnetic, and upper atmosphere and auroral observations was stepped up. The Vestfold Hills are one of the largest of the ice-free Antarctic oases, and Davis station became a focus for studying the unique ecology of its lakes, as well as contributing to the geophysical program.

▼ **MAWSON'S GROWTH**
The green-roofed hut (right) is one of the station's original (1954) buildings; others were added over time: aluminum workshops and sleeping quarters (center) (1960-70s), the red and green buildings (1980s). The large geodesic dome which houses the satellite communication system was built in the 1990s.

OLD AND NEW AT CASEY STATION
Casey station's unique caterpillar-on-stilts design was used to avoid drifts building up by allowing snow to blow harmlessly beneath. It worked well, but its position so close to the shore exposed it to salt spray that weakened the supporting structure, so the new Casey was built higher up the hill.

Both the United States and the Union of Soviet Socialist Republics built stations in the Australian Antarctic Territory for the IGY; the American station, Wilkes, was on the coast, and the five Russian stations included Vostok, the highest and coldest of the Antarctic bases. In 1959 ANARE took over Wilkes station, and in 1962 six men traveled by tractor from Wilkes to Vostok and back, a distance of 3,000 kilometers (1,860 miles)—the most ambitious overland Antarctic journey attempted to that time. They took ice core samples and fired seismic shots to determine the thickness of the ice cap, and also made magnetic and meteorological observations in an area never traversed before—or since. By 1965 Wilkes station was almost buried in snow, so Davis station closed down for four years to release funds to build a new Australian station, named Casey, to replace Wilkes.

Phillip Law

Born in northern Victoria and educated in Melbourne, Phillip Law was appointed senior scientific officer of the newly conceived Australian National Antarctic Research Expeditions (ANARE) in 1947; two years later he became its director, a position he held until 1966. Over that period he made 28 voyages to Antarctica or the sub-Antarctic and developed a technique of "hit and run" exploration that accurately charted over 5,000 kilometers (3,100 miles) of the coast of the Australian Antarctic Territory (AAT). When Law retired after 17 years as Director of ANARE, the main geographic features of the AAT were known and he had established a strong focus on scientific research, both on land and at sea. He had also defined the ANARE style—inventive, resourceful, egalitarian yet highly professional—that is still its distinguishing characteristic.

LAW FIELD BASE IN THE LARSEMANN HILLS
As well as permanent stations, ANARE operates a number of summer bases in areas of particular scientific significance. Scientists live in traditional polar pyramid tents (right) or the unique Australian-designed red Apple huts that are easily transported slung beneath a helicopter.

The International Geophysical Year

HUB OF THE ANTARCTIC
McMurdo station on Ross Island was first established in 1955–56 by Operation Deep-freeze I. It was the advance staging post for building Amundsen–Scott base at the South Pole in 1957, and has since become by far the largest base. Housing up to 2,000 people in summer, its dusty gravel roads give it the ambience of a mining town.

The International Geophysical Year (IGY) focused attention on Antarctic science, and was a stepping stone to the Antarctic Treaty of 1959. The idea originated in a discussion among American and British physicists in 1950, and took its theme from the International Polar Years of 1882–83 and 1932–33. The second had been a period of minimum sunspot activity; 1957–58 would be a period of maximum activity, so findings would provide valuable comparisons. The proposal was endorsed by the International Council of Scientific Unions, which set up a committee in 1952 to organize what became the IGY. Its scientific program focused on two areas, Antarctica and outer space, and ran for 18 months from 1 July 1957.

Twelve nations—Argentina, Australia, Belgium, Britain, Chile, France, Japan, New Zealand, Norway, South Africa, the United States, and the Union of Soviet Socialist Republics—established 50 stations in Antarctica, with a total of more than 5,000 personnel, to conduct a research program focusing on glaciology, meteorology, geology, geomagnetism, and upper atmosphere physics. The largest contributors were the Union of Soviet Socialist Republics, with seven Antarctic stations, including one at the Pole of Inaccessibility (the point on the Southern Continent furthest from all its coasts), and America, also with seven stations, including Amundsen–Scott station at the South Pole.

Paul-Emile Victor

1907–95

Paul-Emile Victor was President of the French Antarctic Committee for the International Geophysical Year. A leading figure in French polar exploration, he organised his first expedition in 1934, and was Director of the French Polar Expeditions from their beginnings in 1947 until 1976. Over many years he conducted numerous expeditions to Greenland in the Arctic, and, from 1949 to 1951, to the permanent French scientific base on Terre Adélie in Antarctica.

In 1987, to celebrate his 80th birthday, Victor returned to Terre Adélie, and then to the North Pole.

Victor died on 7 March 1995. According to his will he was immersed off Bora Bora in French Polynesia, from the National Marine frigate *Dumont D'Urville*.

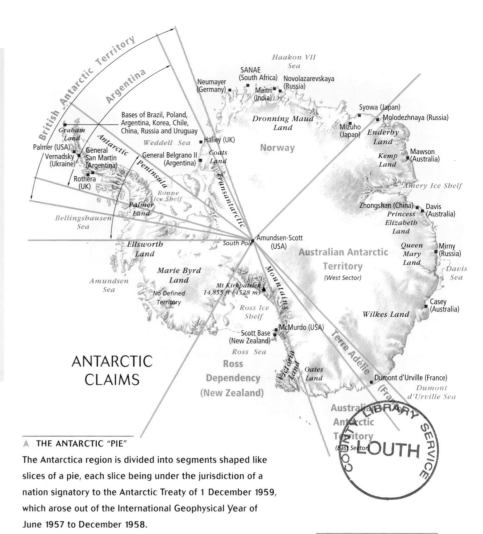

▲ **THE ANTARCTIC "PIE"**
The Antarctica region is divided into segments shaped like slices of a pie, each slice being under the jurisdiction of a nation signatory to the Antarctic Treaty of 1 December 1959, which arose out of the International Geophysical Year of June 1957 to December 1958.

Operation Deepfreeze

This was the four-year program through which the United States met its IGY objectives, under the overall command of Admiral George Dufek. Dufek's task was to coordinate the resources of the United States Navy, Army, Air Force, Marines, and Coast Guard with those of other nations, particularly New Zealand, and to have all seven bases operational, equipped, and staffed with civilian scientists by July 1957—a formidable challenge.

Deepfreeze I (1955–56) established McMurdo station on Ross Island and Little America V on the Ross Ice Shelf. From the bridgehead at Little America, Deepfreeze II (1956–57) brought 12 ships, 13 aircraft, and 4,000 men to undertake the far more difficult task of establishing Amundsen–Scott base at the South Pole, and Byrd base, 1,000 kilometers (650 miles) inland from Little America.

During Deepfreeze III (1957–58), 350 men wintered at the seven United States bases to conduct research. Little America V was Weather Control for the entire Antarctic meteorological network, and an international staff coordinated daily observations. Seismic soundings taken at the South Pole revealed that the ice cap at the Pole was nearly 2,750 meters (9,000 ft) thick, and glaciological traverses helped to delineate the contours of the bedrock far below.

"A continent for peace and science"

Perhaps the IGY's greatest success was that it focused attention on the need for a permanent international regime to manage Antarctica, and prevent it from falling prey to cold war conflict between the superpow-

ers. After 18 months of negotiations, the Antarctic Treaty was signed in December 1959 by the 12 nations that had participated in the IGY. The Treaty banned military activity, guaranteed free access for scientific research, and solved the problem of territorial claims by stating that the Treaty did not "endorse, support or deny any territorial claim," and prohibiting any further claims. Under this regime, Antarctica has become a unique area.   LC

▼ **HOW THICK IS THE ICE?**
Traveling on Ski-Doos (motor toboggans), a pre-season sea ice survey team drills through the sea ice in McMurdo Sound, mapping its thickness. This annual survey gauges the suitability of the McMurdo Sound fast ice for surface travel during the summer.

Wildlife conservation status

Conservation status information from two internationally recognized sources, the World Conservation Union (IUCN) and the Convention on International Trade in Endangered Species of Wild Fauna and Flora (CITES), are included with other statistics at the top of species entries in the wildlife chapter of this book. An explanation of the status codes and the organizations which generate them are outlined below.

IUCN

The World Conservation Monitoring Center (WCMC) is part of the United Nations Environment Program (UNEP), and provides worldwide biodiversity information for policy and action to conserve the living world. That role includes publication of the *IUCN Red List of Threatened Plants and Animals*, which is an internationally recognized series of lists that categorize the status of globally threatened species. The WCMC determines the relative risk of extinction of species, and the main purpose of the Red List is to catalogue the species that are regarded as threatened at a global level and are at risk of overall extinction.

Red List Categories and Criteria

EXTINCT

A taxon is Extinct when there is no reasonable doubt that the last individual has died.

EXTINCT IN THE WILD

A taxon is Extinct in the wild when it is known only to survive in cultivation, in captivity, or as a naturalized population (or populations) well outside its past range.

CRITICALLY ENDANGERED

A taxon is Critically Endangered when it is facing an extremely high risk of extinction in the wild in the immediate future: an 80 percent reduction over 10 years or three generations; at least 50 percent probability of extinction in the wild within 10 years.

ENDANGERED

A taxon is Endangered when it is facing a very high risk of extinction in the wild in the near future: a reduction of at least 50 percent over 10 years or three generations; severely fragmented or reduced populations; at least 20 percent probability of extinction in the wild within 20 years.

VULNERABLE

A taxon is Vulnerable when it is facing a high risk of extinction in the wild in the medium-term future: 20 percent probability of extinction in the wild within 10 years or three generations; very small or restricted population; at least 10 percent probability of extinction in the wild within 100 years.

LOWER RISK

The Lower Risk category is separated into three subcategories:-

CONSERVATION DEPENDENT: Taxa which are the focus of a continuing conservation program, the cessation of which would result in the taxon qualifying for one of the threatened categories above within a period of five years.

NEAR THREATENED: Taxa which do not qualify for Conservation Dependent, but which are close to qualifying for Vulnerable.

LEAST CONCERN: Taxa which do not qualify for Conservation Dependent or Near Threatened.

DATA DEFICIENT

A taxon is Data Deficient when there is inadequate information to make an assessment of its risk of extinction based on its distribution and/or population status. Data Deficient is not a category of threat or Lower Risk, and acknowledges the possibility that future research will show that threatened classification is appropriate.

NOT EVALUATED

A taxon is Not Evaluated when it is has not yet been assessed against the criteria.

CITES

The international wildlife trade has caused massive declines in the numbers of many species of animals and plants. An international treaty, known as CITES, to which 152 nations subscribe, was introduced in 1975 to ban commercial international trade in endangered species and to regulate and monitor trade in other species that might become endangered.

CITES APPENDIX I

The most endangered species are listed in Appendix I, which includes all species threatened with extinction that are or may be affected by trade.

CITES APPENDIX II

Other species at serious risk are listed in Appendix II, which includes all species that although not necessarily currently threatened with extinction may become so unless trade is subject to strict regulation. Appendix II includes other species which must be subject to regulation in order that trade in endangered species listed in Appendix I may be brought under effective control, for example, species similar in appearance to endangered species.

Additional information can be found at the following websites:

http://www.iucn.org/themes/ssc/redlists/ssc-rl-c.htm
http://www.cites.org/CITES/eng/index.shtm

◄ Adélie penguins, Mt Herschel.

Antarctica around the world

Many sites around the world outside Antarctica have an association with the south polar regions, some very directly so, others rather tenuously. They offer insights into the history of the region and an affordable alternative to going to the Antarctic as a tourist. Some of these "low-latitude" Antarctic locations are listed below. Visit http://www.antarctic-circle.org/llag.htm for more.

England

London and its environs

Among the museums and public collections worth visiting, the recently enlarged **National Maritime Museum** at **Greenwich, S.E.10**, has also expanded its Antarctic collection. Along with the Scott Polar Research Institute in Cambridge, this is Britain's largest accumulation of Antarctic memorabilia. It contains numerous paintings from the Cook era onward, the autographed banjo that L.D.A Hussey played on Elephant Island, a Scott sledge, his sledge flag flown at the South Pole, and much more.

The **Royal Geographical Society, 1 Kensington Gore, S.W.7**, is a treasure trove of artefacts, paintings, and manuscripts, though most are stored away. A large model of the *Discovery* sits just inside the entrance, and paintings of James Cook, Robert Scott, and his mentor, Sir Clements Markham, hang nearby (a bust of Markham is set outside, next to the doorway). A powerfully evocative bronze statue of Ernest Shackleton occupies a niche on the building's Exhibition Road façade.

Shackleton's family lived at **12 Westwood Hill, S.E.26**—this house in which he grew up still stands and is marked with a historic blue plaque. As a youngster, Shackleton attended nearby **Dulwich College, S.E.21**. The college has a large collection of "Shackletonia" in its archives, but the star attraction, on show in the North Cloisters, is the *James Caird*, the 7-meter (23-foot) boat that made the incredible 1,287-kilometer (800-mile) voyage from Elephant Island to South Georgia. With raised sails it sits on a bed of stones brought from South Georgia.

The **British Library, 96 Euston Road, N.W.1**, has just one Antarctic item on public display, but it is one that should not be missed: Scott's journal opened to his final entry, written on 17 March 1912.

Houses of Antarctic explorers are scattered about London. Scott rented **56 Oakley Street, S.W.3** (marked by a blue plaque), from 1904 to 1908, and lived with his mother and sisters at **80 Hospital Road, S.W.3**, while organizing the *Discovery* expedition. Shackleton had many addresses over the years, one of the more interesting was **11 Vicarage Gate, Kensington, W.8**, where he moved in 1913; it is now the Abbey House Hotel. Two other blue-plaque Antarctic house are those of Sir James Clark Ross at **2 Eliot Place, Blackheath, S.E.3**, and Dr Edward A. Wilson at **42 Vicarage Crescent, Battersea, S.W.11**.

Statuary and memorials also abound in London and its environs. Some notable ones are the bronze statue of Robert Scott in sledging gear sculpted by his artist wife Kathleen—she did a second version in marble, which may be found in Christchurch, New Zealand. The bronze version, unveiled in 1915, stands in **Waterloo Place, S.W.1**, opposite the statue of another tragic polar hero: Arctic explorer Sir John Franklin. At **St Paul's Cathedral, E.C.4**, on 5 May 1916, Prime Minister Asquith unveiled a large bronze tablet in the south transept in memory of Scott and his four polar companions—the men are depicted hauling a sledge. Above Scott's head is the inscription: "Death is swallowed up in victory."

The bronze plaque in the school library at **Eton College**, near **Windsor**, is another Kathleen Scott work. It memorializes L.E.G. Oates, the "very gallant gentleman" who willingly went to his death to save his comrades.

Elsewhere in England

The **Scott Polar Research Institute (SPRI), Lensfield Road, Cambridge**, is with little doubt the ultimate destination for polar pilgrims. It has long been the world's preeminent polar educational and research institution, but it is the museum, archives, and library that draw enthusiasts, writers, and collectors from near and far. The relics, journals, paintings, and expedition equipment on display are numerous and varied. Some favorite items include five framed Antarctic sledging flags, Captain Scott's "housewife" or repair kit, and Edward Wilson watercolors (only a tiny selection of the SPRI's collection are on show), the wooden box used for anonymous submissions to the *South Polar Times*, Oates' sleeping bag, and his last letter to his mother. Elsewhere in the building are the brass ship's bell from the *Terra Nova* (rung twice a day) and a spar from the *Endurance*.

The **Oates Memorial Museum** in the lovely village of **Selborne, Hampshire**, shares space with the Gilbert White Museum in "The Wakes", where White wrote his famous natural history book. The collection focuses on Captain Oates who died on the return from the South Pole and includes artefacts, paintings, a life ring, and a pennant from the *Terra Nova*, together with various pieces of clothing and equipment.

In a similar vein, the exhibit memorializing Dr Edward Wilson at **the Cheltenham Art Gallery & Museum, Clarence Street, Cheltenham, Gloucestershire**, is a popular one. Among special items there is the prayer book from which, on 12 November 1912, surgeon Edward Atkinson read the service for Scott, Wilson, and "Birdie" Bowers beside the tent in which they died. A short walk away, on The Promenade, is a bronze statue of Dr Wilson, sculpted by Kathleen Scott and reminiscent of the one she did of her husband in London. Wilson was born in Cheltenham at **91 Montpellier Terrace**; the fact is noted in a large inscription on the stone façade of the building, which now offers bed and breakfast accommodation.

Plymouth was Scott's hometown, and his skis and a model of the *Terra Nova* are preserved at the **Plymouth City Museum & Art Gallery, Drake Circus**. A few miles away on **Mount Wise** (really a very small hill) in **Devonport** stands the national memorial to Scott and his companions, a mammoth affair of bronze and granite that incorporates some very powerfully executed scenes and bas-relief portraits. The Scott tragedy gave rise to a tremendous number of memorial tributes throughout the Empire. The one at Mount Wise is notable for its size; others have attributes equally as compelling. The four Scott Party Memorial Windows **at St Peter's Church, Binton, Warwickshire**, are admired for their skilful depiction of the polar journey. Scott's brother-in-law was rector of the church and is buried in the churchyard, his grave marker designed by his sister, Kathleen.

On a wall in **St Mary the Virgin Church at Gestingthorpe, Essex**, hangs a brass plaque in memory of Captain Oates who grew up across the road in the local manor house, **Gestingthorpe Hall**. Placed there a year after his death by his brother officers of the Sixth Inniskilling Dragoons, it still shines brightly; his mother used to polish it every week throughout her long life.

In the pleasant village of **Aston Abbots, Buckinghamshire**, behind **St James the Great Church**, is the tomb of Sir James Clark

Ross who gave his name to the sea, the island, and the ice shelf. Inside the church is a memorial window to him and his wife Ann.

Another grave, earlier still, is that of Edward Bransfield in the **Brighton Extra Mural Cemetery, Brighton, East Sussex.** The recently refurbished raised tomb includes the inscription "the first man to see mainland Antarctica in January 1820" (and if Bransfield was not the first, he was nearly so). Bransfield, who died in obscurity, lived at **11 Clifton Road**, later moving to **61 London Road**; both houses are still standing in Brighton.

Wales
Cardiff
Cardiff has strong connections with the Antarctic. It was the last British port to see *Terra Nova* off and the first to welcome her back. The **Royal Hotel** in **St Mary Street** hosted a farewell banquet for the officers and scientists of Scott's expedition on 13 June 1910. The *Terra Nova* departed two days later. On 17 June 1913, a welcome-back dinner was also held there, though surely a more somber gathering. At the **Welsh Industrial and Maritime Museum, 126 Bute Street, Cardiff Bay**, the *Terra Nova*'s figurehead, a lovely blonde lady, holds reign. In 1916 a large bronze memorial tablet to commemorate Scott's sacrifice was installed by the grand staircase in the **Cardiff City Hall**. Its pictorial detail includes penguins, seals, skis, a sledge, and dogs. The **Clock Tower** rising from the lake in nearby **Roath Park** is a memorial to Scott's polar party in the form of a lighthouse.

Scotland
The *Discovery*, Scott's ship on his first Antarctic expedition, was built in **Dundee**. For years it languished in London but has now returned to its birthplace where it is the centerpiece of the harborside development: **Discovery Point**. The ship is open for visits and the wardroom can be hired for functions. The on-shore exhibits are extensive and very well presented.

Sir Ernest Shackleton lived at **14 South Learmouth Gardens, Edinburgh**, from April 1904 while he was secretary of the Royal Scottish Geographical Society. This building is now Channings, an upmarket hotel. Brother officers of "Birdie" Bowers, who died with Scott, erected a memorial to him at **St Ninian's Church** in **Port Bannatyne** on the **Isle of Bute** about a mile north of the center of **Rothesay**. (An identical one was also installed in **St Thomas Cathedral, Bombay, India**—the oldest English building in Bombay and "since 1718 the center of Bombay's Christian worship.")

Ireland
Most Antarctic sites in Ireland have to do with Shackleton in one way or another. Before the family moved to Croydon, south of London in 1884, they spent four years at **35 Marlborough Road, Dublin**, a handsome brick house that recently received a plaque noting this fact. Sir Ernest was born in 1874 at **Kilkea House** in **County Kildare**, and in nearby **Athy**, at the **Athy Heritage Centre** in the Town Hall, an exhibit on the famous native son includes a sledge used during the *Nimrod* expedition.

To the west, at **Anascaul** on the **Dingle Peninsula, County Kerry**, is Tom Crean's **South Pole Inn**, the pub he opened after his Antarctic exploring days with Scott and Shackleton ended. It has some appropriate Antarctic decorations and a guest book that records the names of hundreds of thirsty Antarcticans. Crean's grave is about a mile away in the very wild though picturesque **Ballinacorty Cemetery**. A monument in the **Town Park, Kinsale, County Cork**, has recently been dedicated to the McCarthy brothers: Mortimer, a member of Scott's *Terra Nova* expedition, and Timothy who accompanied Shackleton in the *James Caird*.

France
Paris
France's two most famous Antarctic explorers are buried in **Paris**. Jules-Sébastian-César Dumont d'Urville, who led the great 1838-40 expedition to the South lies in the **Cimetière Montparnasse**. He, his wife, and son were killed in a fiery railroad crash in 1842. His monument is a large rounded obelisk. North of the Seine, in the **Cimetière Montmartre**, is the mausoleum of Jean-Baptiste Charcot, who also died tragically when he went down with his famous ship *Pourquoi-Pas?* during a gale off Iceland. Both cemeteries are known for their celebrity residents.

Norway
Oslo and its environs
Oslo's best known Antarctic site is the **Fram Museum**, where Nansen and Amundsen's famous ship *Fram* has been enclosed in a large A-frame structure since 1936. There are various displays on the ship, not only relating to Nansen and Amundsen but also to Carsten Borchegrevink. Oscar Wisting, who was at the South Pole with Amundsen, asked to spend a night in his old cabin soon after the museum opened; he was found dead there the next morning. The **Ski Museum**, at the **Holmenkollen ski jump**, north of Oslo, has Amundsen artefacts including a tent, skis, clothing, provision boxes, even a stuffed sledge dog. Amundsen's house, **"Uranienborg,"** is located south of the city, at **Svartskog** on the east side of the fjord and now open as a museum. His bedroom features portholes, and everything is just as he left it when he set out on his ill-fated rescue attempt of Umberto Nobile. Down by the fjord, is a bronze statue of Amundsen, accompanied by a sledge dog.

Elsewhere
The **Bjaaland Museum** in **Morgedal, Telemark,** remembers Olav Bjaaland, the last of the five in Amundsen's polar party to die, in 1961. Oscar Wisting was born in **Larvik** and a statue of him stands outside the **Sjøfartsmuseum**; close by is a bust of Colin Archer, the designer of the *Fram* who also lived in Larvik. A statue of Amundsen graces the public square in **Tromsø** and some Amundsen artefacts are preserved in the **Polar Museum** there. Although there is no monument to Amundsen in Britain, Scott is commemorated in Norway, by a rough-hewn granite obelisk near the railroad station at **Finse** on the line between Oslo and Bergen.

Japan
Lieutenant Nobu Shirase was Japan's participant in the Heroic Age; his team actually met some of Amundsen's party in the Bay of Whales. The **Shirase Antarctic Expedition Memorial Museum** is located in his hometown of **Konoura**, 500 kilometers (310 miles) north of Tokyo. Among the exhibits is a full-scale model of the stern section of his ship, *Kainan Maru*, and equipment and clothing from his expedition.

South Africa
British ships bound for the Antarctic generally re-provisioned in **Cape Town**, so it is not surprising to find a memorial to Scott in **Adderly Street**. The **Johannesburg Municipal Gallery** has a lovely oil portrait of Scott's wife Kathleen by Charles Shannon. Dr Reginald Koettlitz, physician on the *Discovery* expedition, is buried at **Somerset East**.

United States
Washington and its environs
The **Navy Museum** in the **Washington Navy Yard** is a cavernous building filled with exhibits including a whole section on polar exploration, the largest of its kind in the United States. Byrd, Ronne, Wilkes, and other Americans are highlighted, but attention is also paid to Scott, Shackleton, and Dumont D'Urville. Lincoln Ellsworth's *Polar Star*, the plane in which he and Hollick-Kenyon made the first traverse of Antarctica, can be found at the Smithsonian's **National Air and Space**

Museum on the Mall. A plaque beside a footpath just north of the Capitol notes where once Lieutenant Charles Wilkes' house stood. Not too far away, on Pennsylvania Avenue at Eighth Street, is the elaborate Navy Memorial. Among its many bas-reliefs is one devoted to Wilkes. His grave is in Arlington National Cemetery across the Potomac, and the marker states, incorrectly, that "he discovered the Antarctic Continent." Three other Antarctic graves are there: that of Admiral Richard Byrd with a simple military issue headstone, behind it the far larger marker for Bernt Balchen, and nearby that of Finn Ronne. A large bronze statue of Byrd stands on the Avenue of Heroes across from the Visitors Center.

New England

Although Byrd was from an old Virginia family, he married a Boston woman and when not exploring in the north or south was at home at 9 Brimmer Street, a lovely brick townhouse at the foot of Boston's Beacon Hill. Igloo, the fox terrier that accompanied Byrd on his early Antarctic travels, is buried in the Pine Ridge Cemetery for Small Animals in Dedham, just outside Boston. The iceberg-shaped, pinkish gravestone is inscribed "Igloo. He was more than a friend."

The splendid Peabody Essex Museum in Salem, Massachusetts, has a number of Antarctic artefacts though few are displayed. However, Charles Wilkes' superb oil painting of his flagship *Vincennes* hangs in the East India Marine Hall.

Many of the early sealers in the South Shetlands called Stonington, Connecticut, their home port. Nathaniel Brown Palmer—who along with Bellingshausen, Smith, and Bransfield, claimed to be the first to sight the Antarctic continent in 1820—is buried in the tranquil Evergreen Cemetery. The grave of Phineas Wilcox, the first mate of Palmer's *Hero*, is in the Miner Burying Ground nearby. A sign marks Captain Palmer's birthplace on Water Street and the Stonington Historical Society headquarters now occupies his ornate mansion on Palmer Street. Within are various paintings, relics, and displays relating to Palmer, the Antarctic, and sealing.

Elsewhere

The enormous American Museum of Natural History in New York City contains a ground-floor alcove display that focuses on Lincoln Ellsworth—a Museum trustee—but also includes many interesting items associated with Amundsen and Nordenskjöld. In Annapolis, Maryland, the excellent Naval Academy Museum has, among other things,

Wilkes' dress cocked hat, and sledges associated with both Byrd and Scott. The Virginia Aviation Museum at the Richmond International Airport is home to Byrd's aircraft, *Stars & Stripes*. Another famous Byrd aircraft, the Ford tri-motor *Floyd Bennett*—first to *fly over* the South Pole—is at the Henry Ford Museum in Dearborn, Michigan. The *Que Sera Sera*—first plane to *land* at the South Pole—is at the National Museum of Naval Aviation in Pensacola, Florida.

Ohio State University is home to the Byrd Polar Research Center. A few items are on display, but this is primarily a research and educational center, the closest thing to an American SPRI. The University Archives nearby house the papers of Byrd, Sir Hubert Wilkins, and Dr Frederick Cook. Ohio is also the location of a remarkable monument to John Cleves Symmes. He originated the Theory of Concentric Spheres and Polar Voids, or simply the Hollow Earth Theory. Appropriately, his monument on Third Street in Hamilton is a stone obelisk, atop which sits a round stone sphere with a hole drilled through it.

Argentina

Argentina is a likely jumping-off place for those traveling to the Antarctic Peninsula, and there are several sites to visit on the way. In Buenos Aires, the *Uruguay*, a famous Antarctic ship, is now a museum moored at Dock One along the lengthy Perto Madero. Far to the south, in Ushuaia, the Maritime Museum is located in the Presidio. It has the largest collection of same-scale Antarctic ship models anywhere, numerous artefacts, particularly ones associated with Nordenskjöld's expedition, and an extensive philatelic display.

Chile

Santiago's Natural History Museum has dioramas of Antarctic bird life, archeological items and relics from the South Shetlands, and displays of polar equipment.

Australia

Tasmania: Hobart

Hobart has traditionally been the departing point for voyages to the Antarctic from Australia; among the earliest were Sir James Clark Ross and Dumont d'Urville. At the Hobart Crematorium and Cemetery, Cornellian Bay, is a memorial to the sailors on Dumont d'Urville's expedition who died at sea or in hospital in Hobart and a commemorative rose garden. There is some excellent material on sealing, whaling, and Antarctic expeditions leaving from Tasmania at the State Library of Tasmania in Murray Street; early

whaling gear and Antarctic expedition equipment may be seen at the Tasmanian Museum and Art Gallery in Macquarie Street. In St Albans Anglican Church, Main Road, Claremont, are stained glass windows memorializing the Scott expedition.

There is a small museum of Heroic Age artefacts in the foyer of the Australian Antarctic Division, Channel Highway, Kingston, and the Division also boasts world-class collections of polar library materials and Antarctic image resources. A bust of Amundsen may be found in the Centenary Building at the University of Tasmania Sandy Bay campus. The same building is home to two special Antarctic-related organizations: the Institute of Antarctic and Southern Ocean Studies (IASOS) and the prestigious Antarctic Cooperative Research Centre (CRC).

One can spend a night in the Amundsen Suite at Hadley's Hotel in Murray Street—Amundsen was a guest there in March of 1912 on his return from the Pole. During his short stay he sent a telegram from the Hobart General Post Office on the corner of Macquarie and Elizabeth Streets, informing the world of his triumph. "Antarctic Adventure", Salamanca Square, presents hands-on exhibits. Hobart's Maritime Place was recently renamed Mawson Place in honor of the great explorer; the associated Waterside Pavilion features an Antarctic display.

New South Wales: Sydney and Newcastle

Among its vast collection of Antarctic documents and photographs, Sydney's Mitchell Library holds Apsley Cherry-Garrard's Antarctic sketchbooks and Frank Hurley's 20 surviving Paget color transparencies from the *Endurance* expedition. The Powerhouse Museum, Darling Harbour, Sydney, has several items associated with Mawson including a sledge, clothing, snow goggles, and Charles Laseron's polar medal. "In Blizzard Bound," an exhibit at the Newcastle Regional Museum, also features items from Mawson's expeditions.

ACT: Canberra

Mawson, a suburb of Canberra, was established in 1966. The streets were named for persons and ships associated with the Antarctic, both Australian and otherwise, including Colbeck, Hurley, Joyce, Markham, Shackleton, and Wilkins.

Victoria: Melbourne

Captain Cook's cottage is a popular tourist spot in a lovely setting in Fitzroy Gardens, Melbourne. It was moved there from Great

Ayton, North Yorkshire, in 1934. (The world is full of Cook sites—**North Yorkshire, England,** being particularly rich in them.) In **St Paul's Cathedral** a wood plaque was recently installed honoring Australian Antarctic expeditioners who made the ultimate sacrifice. Outside the **Polly Woodside Maritime Museum, Southbank,** is a memorial to the Antarctic ship *Nella Dan*, while inside the Museum is a large model of the ship and a "Secrets of the Frozen World" exhibit.

South Australia: Adelaide

A bronze bust of Australia's great Antarctic explorer, Sir Douglas Mawson, stands in an outdoor setting at the **University of Adelaide** where he taught for nearly 50 years. Although the University's Mawson Institute closed in 1990, there is a display of Mawson memorabilia—including the famous "half-sledge"—at the **Tate Geological Museum.** At the **Waite Campus,** hundreds of Mawson objects are in storage, awaiting a better and more accessible venue, an effort also involving the **South Australian Museum, North Terrace,** which also has many Mawson items.

Western Australia: Esperance

Charles Sandell brought back souvenirs and relics from Mawson's Australasian Antarctic Expedition, and donated them to the **Esperance Municipal Museum** in his hometown: among them is a propeller from the "air tractor" and a 1911 Christmas bottle of port.

New Zealand

New Zealand has had a particularly close association with Antarctic expeditions from the Heroic Age to the present, and many sites there are worth visiting.

Christchurch and its environs

The **Visitor Centre** at the **International Antarctic Centre** beside **Christchurch Airport** provides an interactive experience focusing on Antarctic science and the polar environment, a variety of displays including some Heroic Age artefacts, and a gift shop and café. The **Canterbury Museum, Rolleston Avenue, central Christchurch,** shares with the SPRI in England the title of world's premier Antarctic museum. Some favorite items include Frank Hurley's Model B Kodak camera, the stove that accompanied Shackleton on the *James Caird* and the silver communion service from Scott's Cape Evans hut. Close by in the port of **Lyttelton** is the **Lyttelton Museum** on Gladstone Quay. This eclectic local history museum has an Antarctic component including a stuffed Emperor penguin, polar medals, a model of the *Discovery*, and a Shackleton sledge. In the center of **Christchurch** on **Worcester Street** stands the statue of Captain Scott sculpted by his wife Kathleen, a copy of the bronze one in London. Because of wartime metal shortages Italian marble was used. Two plaques with Antarctic connections can be found at **Christchurch Cathedral.**

Auckland and its environs

The **Auckland Museum** in the **Domain** has a small display of Heroic Age memorabilia. **Kelly Tarlton's Antarctic Encounter** is a theme park of sorts (offering rides in Snow Cats) about four miles from the center of town. Among other things, particularly penguins, it features a replica of the wardroom in Scott's hut at Cape Evans.

Wellington and its environs

The **Alexander Turnbull Library** is New Zealand's national library and holds many Antarctic-related manuscripts and journals, as well as two copies of the very rare *Aurora Australis,* issued at Cape Royds during Shackleton's *Nimrod* expedition. **Mount Victoria** rises above the city and on top is a memorial to **Richard Byrd.** Shaped like a polar tent, facing south, it was dedicated in 1962 and extensively repaired a few years ago. The graves of two important Antarcticans associated with Shackleton's *Endurance* expedition are in **Karori Cemetery**: Harry "Chips" McNeish, the carpenter who kept the *James Caird* together, and Colonel Thomas Orde-Lees.

Port Chalmers and Dunedin

The departure of Scott's *Discovery* from Lyttelton in 1901 was marred by the death of Charles Bonner who fell from the top of the mainmast when the ship encountered an ocean swell. He was buried at the next port of call in the **Port Chalmers Cemetery** where a stone obelisk marks the grave. The most impressive monument in **Port Chalmers**— given size and setting—is the **Scott Memorial,** a circular 9-meter (30-foot) tall stone column with an anchor on top, set high on a hilltop overlooking the harbor. A bronze bust of Richard Byrd can be found in **Unity Park, Dunedin**; it was presented to the city by the National Geographic Society.

Elsewhere

A memorial to Scott and his companions is situated in the park beside **Lake Wakatipu** in **Queenstown.** It consists of two plaques—one with Scott's last message, the other with the names of the polar party—set on a huge glacial boulder. At **Waitaki Boys' High School** in **Oamaru** "Birdie" Bowers's sledge flag, made by his sister, is in the Hall of Memories, which also contains a tablet in Scott's memory. ROBERT B. STEPHENSON

Gazetteer

- The gazetteer contains the names on the following maps: Antarctica (pages 8–9), Antarctic Peninsula (page 57), Gerlache Strait (page 59), Ross Sea and Ross Island (page 63), Antarctica and the Sub-Antarctic Islands (pages 68–69).
- The country or ocean is listed for all names outside Antarctica and the Sub-Antarctic islands.
- Names that have a symbol (stations and summits) are given a grid reference in the gazetteer according to the location of the symbol. Names without a symbol are referenced in the gazetteer according to the start or center of the name.

Index

Page numbers in *italics* refer to photographs.

Acknowledgments

The authors would like to thank the many people who have contributed to this project. Primarily, we are grateful to all the contributors:

The late Don Adamson of Macquarie University in Sydney, Australia, who, over 23 years, undertook summer fieldwork in continental Antarctica and on sub-Antarctic Macquarie Island in geomorphology, vegetation–landscape interactions, and environmental history.

Dr Gary Burns, a Principal Research Scientist at the Australian Antarctic Division who has wintered at Casey and on Macquarie Island, and spent a summer at Davis.

Louise Crossley, the author of two books on Antarctica and who has spent two winters and three summers in Antarctica with ANARE as well as serving as a guide/lecturer on icebreaker cruises to the Antarctic Peninsula, the Weddell Sea, and the Ross Sea.

Dr Arthur Ford who, for nearly three decades, participated in numerous United States Geological Survey expeditions into many areas of Antarctica and is a recipient of the United States Antarctica Service Medal. The chapter on Antarctica in *Encyclopædia Britannica* is among his more than 200 publications.

Professor Robert Hill spent 20 years researching the plant macrofossil record in Tasmania and Antarctica over the past 40 million years. He is Professor Emeritus at the University of Tasmania and Professor at the University of Adelaide.

Paul Holper is Communication Manager for Australia's CSIRO Atmospheric Research division. Paul is the author of numerous articles on science and the environment, five science textbooks, and a series of science books for children.

Dr Paul Lehmann has worked in upper atmospheric physics at the University of Melbourne, the University of Illinois and the Max Planck Institute, Germany. In 1989 Paul became a senior physicist at the Bureau of Meteorology, Melbourne and has represented Australia at UN meetings on ozone depletion.

Professor Harvey Marchant is Leader of the Australian Antarctic Biology Program at the Australian Antarctic Division and a Professorial Fellow of the University of Melbourne. He is editor of the international journal, *Polar Biology*.

Dr Gary Miller is a Research Assistant Professor of Biology at the University of New Mexico and specializes in the behavior, ecology, genetics, and diseases of penguins and skuas.

Gary has participated in United States', New Zealand, and Australian Antarctic programs.

Dr Stephen R. Rintoul studies the impact of Southern Ocean currents on the earth's climate. He leads the Southern Ocean program at CSIRO Marine Research.

Dr Pat Selkirk is a Senior Lecturer in biology at Macquarie University and over the past 22 years has undertaken fieldwork on Ross Island, at Casey, and on Macquarie and Heard islands researching plant–environment interactions, environmental history, biogeography, and evolutionary biology.

The authors would also like to extend thanks to the dedicated production team and all at Global Book Publishing. We owe a lot to those with whom we have traveled to the polar regions. Among the many Antarcticans we would like to thank, in no particular order, are Thomas Bauer, Chris Downes, Margot Morrell, Andrew Prossin, the Wikander family, Carla Santos, Greg Mortimer, Andie Smithies, Graeme Watt, Emily Slatten, Ingrid Visser, Robert Clancy, Rinie van Meurs, Dava Sobel, Peter Lemon, Ben Hodges, Lee Belbin, Peter Gill, Nigel de Winser, Francis Herbert, Julia Green, Sir Ed Hillary and June Mulgrew, Phil Law, Stephen Nicol, Glenn A. Baker, Dean Peterson, Rob McNaught, Rinie Van Meurs, Chris McKay, Esteban de Salas, John Borthwick, Dennis Collaton, Chris Doughty, Kim Pitt, Tony Press, Kerry Lorimer, Adrian Truss, Bernhard Lettau, Glenn Browning, Michael Stoddart, Pat Quilty, Aaron & Cathy Lawton, Rob Harcourt, Annie Rushton, Austin Simpson, Kim Westerskov, Pamela Wright, Scott MacPhail, Katie Weekes, Andrew Fountain, Michael Bryden, Zaz Shackleton, Pat Toomey, Jayne Paramor, Peter Hillary, Bill Davis, Des Cooper, Hugh Pennington, Jack Sayers, Darrel Schoeling, Robert Headland, Barry Griffiths, Dave Briscoe, Denise Landau, Heather Jeffery, Dick Filby, Sean Stephen, Mark Eldridge, Tim Bowden, Rob Stephenson, Ken Morton, Andy Beatty, Bernardette Hince, Lincoln Hall, Stephen Martin, Warren Papworth, Adelie Hurley, Toni Hurley-Mooy, Bernard Stonehouse, Craig Bowen, Martin Riddle, Barry Boyce, Bob Finch, Tony Harrington, John Palmer and Ross Brewer. There are many others in the Antarctic community who helped a lot, not least of all our Russian captains and their crews. Several institutes were of inestimable assistance including the Scott Polar Research Institute, Oficina Antártida Ushuaia, Australian Geographic Society, Royal Geographical Society, Cambridge University Library, State Library of NSW, Library of the Australian Antarctic Division, US Naval Library.

The publishers would like to thank the following people for their assistance in the production of this book: Louise Buchanan, Robert Clancy, Vanessa Finney, Nick Gales, Peter Gill, Mark Hindell, David Howard, Harvey Marchant, Gary Miller, Therese Potma, Patrick Quilty, Patrick Toomey.

Photographers: David McGonigal, Kim Westerskov, Grant Dixon, Mike Craven, Tony Palliser, Luke Saffigna, Malcolm Ludgate, Craig Potton, Stewart Campbell, Kevin Deacon, Debra Glasgow, Peter Gill, Albert Kuhnigk, Robyn Stewart, Garry Phillips, Harvey Marchant, Steve Nicol, André Martin, Graham Robertson, Patrick Toomey, Jane Francis, J. Howe, R. Hunt, Geoff Longford, Glen A. Baker, and David Keith Jones.

Global Book Publishing would be pleased to hear from photographers wishing to supply photographs.

The Publisher believes that permission for use of the historical and satellite images in this publication, listed below, has been correctly obtained, however if any errors or omissions have occurred, Global Book Publishing would be pleased to hear from copyright owners.

Australian Picture Library/Corbis: 199 bottom left, 199 bottom right, 199 top right.
Hakluyt Society: 158.
NASA: 14 top left, 51.
National Library of Australia: 198 bottom center, 198 top center.

Main front cover photograph: Cape Royds, Ross Sea.
Front cover inset photograph: Ice caves, Vestfold Hills.
Back cover photograph: King penguins.

pp 12–13: Mount Erebus, Ross Island.
pp 54–55: Ross Ice Shelf.
pp 74–75: King penguin colony, South Georgia.
pp 144–45: McMurdo Sound, near Cape Bernacchi.